Terrorism, Security, and Computation

Series Editor

V. S. Subrahmanian, Department of Computer Science and Institute for Security, Technology and Society, Dartmouth College, Hanover, USA

The purpose of the Computation and International Security book series is to establish the state of the art and set the course for future research in computational approaches to international security. The scope of this series is broad and aims to look at computational research that addresses topics in counter-terrorism, counter-drug, transnational crime, homeland security, cyber-crime, public policy, international conflict, and stability of nations. Computational research areas that interact with these topics include (but are not restricted to) research in databases, machine learning, data mining, planning, artificial intelligence, operations research, mathematics, network analysis, social networks, computer vision, computer security, biometrics, forecasting, and statistical modeling. The series serves as a central source of reference for information and communications technology that addresses topics related to international security. The series aims to publish thorough and cohesive studies on specific topics in international security that have a computational and/or mathematical theme, as well as works that are larger in scope than survey articles and that will contain more detailed background information. The series also provides a single point of coverage of advanced and timely topics and a forum for topics that may not have reached a level of maturity to warrant a comprehensive textbook.

Laura Mostert • Roy Lindelauf
Chiara Pulice • Marnix Provoost
Priyanka Amin • V. S. Subrahmanian

Machine Learning Techniques to Predict Terrorist Attacks

Exemplified by Jama'at Nasr al-Islam wal Muslimin

Laura Mostert
Breda, The Netherlands

Chiara Pulice
Concord, NH, USA

Priyanka Amin
East Meadow, NY, USA

Roy Lindelauf
Breda, The Netherlands

Marnix Provoost
Breda, The Netherlands

V. S. Subrahmanian
Evanston, IL, USA

ISSN 2197-8778　　　　　　ISSN 2197-8786　(electronic)
Terrorism, Security, and Computation
ISBN 978-3-031-93173-4　　　ISBN 978-3-031-93174-1　(eBook)
https://doi.org/10.1007/978-3-031-93174-1

© The Editor(s) (if applicable) and The Author(s), under exclusive license to Springer Nature Switzerland AG 2025

This work is subject to copyright. All rights are solely and exclusively licensed by the Publisher, whether the whole or part of the material is concerned, specifically the rights of translation, reprinting, reuse of illustrations, recitation, broadcasting, reproduction on microfilms or in any other physical way, and transmission or information storage and retrieval, electronic adaptation, computer software, or by similar or dissimilar methodology now known or hereafter developed.
The use of general descriptive names, registered names, trademarks, service marks, etc. in this publication does not imply, even in the absence of a specific statement, that such names are exempt from the relevant protective laws and regulations and therefore free for general use.
The publisher, the authors and the editors are safe to assume that the advice and information in this book are believed to be true and accurate at the date of publication. Neither the publisher nor the authors or the editors give a warranty, expressed or implied, with respect to the material contained herein or for any errors or omissions that may have been made. The publisher remains neutral with regard to jurisdictional claims in published maps and institutional affiliations.

This Springer imprint is published by the registered company Springer Nature Switzerland AG
The registered company address is: Gewerbestrasse 11, 6330 Cham, Switzerland

If disposing of this product, please recycle the paper.

Foreword

Mali is an enormous country in the Sahel region in Africa. In recent years, it has been mostly known for the Tuareg uprising in 2012 and the United Nations Multidimensional Integrated Stabilization Mission in Mali (MINUSMA), which was deployed in 2013 to support the Malian government in restoring stability. The mission ended at the end of 2023, as requested by the Malian government.

As Force Commander in MINUSMA, I came to know the Malian people as very proud and resilient. Men, women, and families with children live their lives in harsh conditions. I have also seen beautiful places, such as the historic Djinguereber Mosque in Timbuktu, built in 1327, or the Bandiagara Cliffs. This 150-kilometer-long sandstone escarpment has served as home to the Dogon people, who are believed to be one of the oldest surviving African cultures. These are beautiful places that normally welcome thousands of tourists, but unfortunately, since 2012, hardly any.

I have experienced the harshness in the northern part of Mali, where infrastructure and basic conditions are lacking. The effects of climate change have made it even more difficult for farmers to grow crops and for herders to take care of their livestock. Unfortunately, the instability has brought insecurity and jihadism on a large scale, affecting big parts of the country.

The situation in Mali is extremely complex. It was the most complex environment I have ever faced in any of my deployments. It was not just the conflict in the north between the Tuareg people and the Malian government. There were more layers of conflict. Jihadist-motivated groups have been able to gain control over vast areas in the northern and central parts of Mali. Jama'at Nasr Al-Islam Wal Muslimin (JNIM) is one of them, and the Islamic State in the Greater Sahel (ISGS) is the other one. It is a mistake to think both are well-organized, smoothly functioning organizations. Lots of subgroups, although affiliated, operate quite autonomously. In addition, intercommunity violence is complicating things further. Finally, there is still illegal trafficking, for which Mali is known throughout its history.

When operating in a conflict-affected area, situational understanding is crucially important. We need it if we want to outmaneuver opposing groups' intentions. It is about knowing *why* things are happening and grasping the opponents' intentions, more than just situational awareness, which is about knowing *what* is happening. Those are the challenges for military intelligence experts. Their work has become more challenging in the past decades.

Apart from the complexity of the operating environment, challenges also come from the amount of information available, the types of actors, the speed of interaction, the use of the information space, including the use of misinformation and disinformation, and geopolitical influences. The challenge for intelligence specialists is to translate the conclusions from all of that into concrete building blocks. That is what a commander needs.

In addition, it is essential to realize that fighting an insurgency is not just a military business. We need the military as part of it, but civilian efforts are necessary to improve governance, provide basic services, and build indigenous capacity. Military efforts need to support and facilitate those efforts. Therefore, coordination with non-military actors, including in the intelligence domain, is essential.

Intelligence is like a puzzle. Putting small pieces together generates a picture. The "so what" question followed by the "now what" question should lead to the right conclusions. Anything that can help put the picture together is helpful. That is why artificial intelligence models will absolutely be helpful in processing the immense amount of information that may be available.

However, the environment is not always as logical as we may think. That is why we will always need the combination of technology and human brains. Machine learning-based models can process lots of information, but let us not only depend on those for conclusions. We will still need human brains to draw and assess conclusions. Information always needs to be put in a context. JNIM is an umbrella organization containing a variety of subgroups, each with its own interests, while intercommunity conflicts, power struggles, opportunistic factors, and ego-styles of behavior will all have their influence as well.

I sincerely compliment the authors for their excellent study in this book. I also commend their conclusions that the knowledge and experiences of subject matter experts are crucial to assure outcomes actionable for a decision-maker. As a former UN Force Commander, I can only strongly confirm that need. As a commander, you always want to be sure that the building blocks that you use as a starting point for your decisions are the right ones. Using a fragile building block can have immense second- and third-order effects that you want to avoid.

Intelligence work will remain challenging and complex. I sincerely hope this book will help us meet those challenges in the right way, using technology and human brains as complements to each other.

Lieutenant-general (retired)
Deputy Commander (ret.) of the Royal Army (Netherlands)

C. J. (Kees) Matthijssen
Force Commander (ret.) United Nations Multidimensional Integrated Stabilization Mission in Mali

Preface

In recent years, the security situation in the Sahel has rapidly deteriorated. Given the Netherlands' prior engagement in the MINUSMA mission, this region captured the interest of the first two authors of this book, recognizing the need for a data-driven approach to understanding the complexities of the region. Therefore, they sought collaboration with Prof. V.S. Subrahmanian and his team, experts in applying machine learning models to analyze terrorist organizations (such as Boko Haram, the Indian Mujahideen, and Lashkar-e-Taiba).

The integration of data science and artificial intelligence (AI) into military operations is no longer a luxury but a necessity. As modern conflicts grow increasingly complex, so too does the need for intelligent, data-driven decision support systems. Recognizing this imperative, the Data Science Centre of Excellence of the Netherlands Ministry of Defence, the Netherlands Defence Academy, and the Northwestern Security and AI Lab have been at the forefront of academic research into the use of AI in military operations. Through collaborative efforts with international ecosystems of academia, government, and industry, these organizations seek to develop and refine methodologies that enhance operational effectiveness and strategic foresight.

This book is the result of these collaborative efforts, specifically addressing the security challenges in the Sahel region. By employing machine learning and advanced analytics, we aim to provide actionable insights that contribute to a more nuanced understanding of the threats posed by non-state armed groups.

This work would not have been possible without the dedication and expertise of many individuals. We extend our deepest gratitude to our research partners, colleagues, and contributors who have provided invaluable insights and support. In particular, we would like to express our sincere thanks to Maaike Tran, Daphne de Vos, and Siebren van der Werf, participants from Defensity College, whose efforts in data gathering and visualization were instrumental in shaping this study. Their diligence and commitment have greatly enhanced the quality of this research, demonstrating the power of interdisciplinary collaboration.

We hope that this book serves as both a foundation and an inspiration for further research into the application of data science in military operations. By leveraging the strengths of AI and analytics, we move closer to a future where data-driven decision-making enhances both operational effectiveness and strategic resilience.

Breda, The Netherlands	Laura Mostert
Breda, The Netherlands	Roy Lindelauf
Concord, NH, USA	Chiara Pulice
Breda, The Netherlands	Marnix Provoost
East Meadow, NY, USA	Priyanka Amin
Evanston, IL, USA	V. S. Subrahmanian
February 14, 2025	

Contents

1 **Introduction**... 1
   1.1 Organization of this Book 4
   1.2 How to Read this Book 6
   1.3 Summary Statistics about JNIM's Violent Activities............ 7
   1.4 Summary of Significant Temporal Probabilistic (TP) Rules 8
   1.5 Efficacy of our Predictive Modeling & the NTEWS System 10
   1.6 Implications for Military Decision Making.................. 18
   1.7 Conclusion.. 18
   References... 19

2 **History of Jama'at Nasr al-Islam wal Muslimin (JNIM)** 21
   2.1 Introduction 21
   2.2 The Sahara-Sahel Region................................ 23
   2.3 JNIM's Origins, Goals, and Strategy 25
   2.4 JNIMs Embedding Within AQ........................... 26
   2.5 JNIMs Area of Operations.............................. 28
   2.6 JNIM as an Insurgent Armed Group 30
   2.7 Conclusion.. 33
   References... 34

3 **Temporal Probabilistic Rules and Policy Computation Algorithms** ... 37
   3.1 JNIM Data 38
   3.2 Temporal Probabilistic Rules............................ 39
   3.3 Conclusion.. 42
   References... 42

4 **Abduction and Release of Abductees** 45
   4.1 Release of Abductees when the National Government Ordered
        Executions and there Was no State of Emergency.............. 49
   4.2 Abductions When a Travel Ban Was Placed on the State
        Where JNIM Operates [AR] 49

	4.3	Release of Abductees When a Travel Ban Was Placed on the State Where JNIM Operates and the Government Did Not Raid Facilities or Institutions [AR]	50
	4.4	Abductions When a Travel Ban Was Placed on the State Where JNIM Operates and JNIM Was Not in Internal Conflict [AR]	51
	4.5	Abductions When Foreign States or International Institutions Froze Assets [AR]	51
	4.6	Abductions When Foreign States or International Institutions Froze Assets and the Government Did Not Raid Facilities or Institutions[AR]	52
	4.7	Abduction When Foreign States or International Institutions Froze Assets and Security Forces Were Not Deserting [AR]	53
	4.8	Conclusions	53
	4.9	Predictive Model/Reports Results	54
	References		55
5	**Attacks on and Targeting of Public Sites**		57
	5.1	Attacks on Public Sites and Travel Bans	62
	5.2	Attacks on Public Sites when Assets Are Frozen	62
	5.3	Attacks on Public Sites When Assets Are Frozen and the Group Was Not in Direct Negotiations	63
	5.4	Attacks on Public Sites When There Is a Travel Ban and the Leadership of the Group Is Not Elected	64
	5.5	Attacks on Public Sites When There Is a Travel Ban and the Group Did Not Communicate through Print Media	64
	5.6	Attacks on Public Sites when their Assets Are Frozen and the Top Leaders of JNIM Are Not under Arrest or Imprisoned	65
	5.7	Attacks on Public Sites when there Is a Travel Ban and the Top Leaders of the Group Did Not Die	65
	5.8	Attacks on Public Sites when Foreign States or Institutions Froze Assets of JNIM and the Government Did Not Warrant Arrests for Members of the Group	66
	5.9	Attacks on Public Sites When a Travel Ban Is Placed on the State Were JNIM Operates and the Group Did Not Communicate a Message of Solidarity	67
	5.10	Attacks on Public Sites When a Travel Ban Is Placed on the State Were JNIM Operates and the Government Did Not Raid Facilities and Locations of the Group	67
	5.11	Conclusions	68
	5.12	Predictive Model/Reports Results	69
	References		70

6	**Targeting of Security Professionals or Security Installations**.......	71
	6.1 No Targeting of Security Forces When There Is No Travel Ban Imposed on the Country Where JNIM Operates and JNIM Did Not Discuss Their Strategy	75
	6.2 No Targeting of Security Forces When Foreign States or International Institutions Did Not Freeze Assets and JNIM Did Not Discuss Their Campaign	76
	6.3 No Targeting of Security Forces When Foreign States or International Institutions Did Not Freeze Assets and JNIM Did Not Issue Claims of Responsibility	76
	6.4 No Targeting of Security Forces When the National Government Security Forces Did Not Execute Civilians and the Government Did Not Declare a State of Emergency	77
	6.5 No Targeting of Security Forces When JNIM Did Not Discuss Its Campaign and the National Government Did Not Declare a State of Emergency	78
	6.6 Conclusions	79
	6.7 Predictive Model/Reports Results	80
	References...	81
7	**Targeting of Civilians**	83
	7.1 Targeting of Civilians When the National Government Instituted a Travel Ban.................................	86
	7.2 Targeting of Civilians When Foreign States or International Institutions Freeze Assets..................................	87
	7.3 Targeting of Civilians When Foreign States or International Institutions Freeze Assets and Top Leaders of JNIM Are Not Under Arrest	87
	7.4 Targeting of Civilians When Foreign States or International Institutions Freeze Assets and Members of JNIM Are Not Under Arrest..	88
	7.5 Targeting of Civilians When Foreign States or International Institutions Freeze Assets and the National Government Did Not Raid JNIM Facilities or Locations...................	89
	7.6 Targeting of Civilians When Foreign States or International Institutions Freeze Assets and the National Government Received Foreign Military Aid.........................	90
	7.7 Conclusions	90
	7.8 Predictive Model/Reports Results	91
	References...	92
8	**Other Types of Attacks**..	93
	8.1 Hit and Run Attacks when the National Government Receives International Military Aid	96
	8.2 Hit and Run Attacks when the National Government Receives International Military Aid and Government Security Forces Did Not Use Sexual Violence	96

	8.3	Hit and Run Attacks when the National Government Receives International Military Aid and the Leadership of JNIM Was Not Split or Fractured	97
	8.4	Sabotage When a Travel Ban Was Placed on the State Where JNIM Operates and there Was no State of Emergency	98
	8.5	Targeting of Public Transportation Facilities When JNIM Addresses the General Public and a Travel Ban Was Placed on the State Where JNIM Operates	98
	8.6	Conclusions	99
	8.7	Predictive Model/Reports Results	100
	References...		101
9	**Reflections & Implications for Military Decision Making**		103
	9.1	The Model-Based Intelligence Cycle........................	104
	9.2	The Value of Adding Geo-Spatial Data	111
	9.3	Defining Variables and Mao's Framework...................	112
	9.4	Conclusion	115
	References...		116
Appendices...			117

Abbreviations

Al-Qaeda (AQ)	A transnational Sunni jihadist non-state armed group pursuing the establishment of a caliphate ruled by sharia by acting as an umbrella organization for regional affiliated non-state armed groups
Al-Qaeda in the Arabian Peninsula (AQAP)	A jihadist non-state armed group on the Arabian peninsula affiliated with Al-Qaeda
Al-Qaeda in the Lands of the Islamic Maghreb (AQIM)	A North African jihadist non-state armed group affiliated with Al-Qaeda
Ansaroul Islam (AI)	A jihadist non-state armed group originating from Burkina Faso, active in the Sahel, and often associated with JNIM's activities
Coordination of Movements of Azawad (CMA)	A coalition of Tuareg and Arab movements advocating for autonomy in northern Mali
Foreign Terrorist Organization (FTO)	A designation by governments to classify groups as engaged in terrorism
Group to Support Islam and Muslims (GSIM)	The English translation of JNIM's name
Human Intelligence (HUMINT)	Intelligence gathered from human sources, crucial for understanding insurgent activities
Improvised Explosive Device (IED)	A type of unconventional explosive weapon that can take any form and be activated in a variety of ways

Islamic State (IS)	A transnational Sunni jihadist non-state armed group competing with Al-Qaeda in the establishment of a caliphate ruled by sharia
Islamic State in the Greater Sahara (ISGS)	A jihadist group competing with JNIM for influence in the Sahel, affiliated with Islamic State (IS)
Jama'at Nasr al-Islam wal Muslimin (JNIM)	A coalition of jihadist non-state armed groups in the Sahel, formed in 2017, with roots in several earlier jihadist groups
Joint Force of the Group of Five for the Sahel (G5 Sahel)	A regional security initiative involving five Sahelian countries to counter terrorism and promote stability
Malian Armed Forces (FAMA)	The national armed forces of Mali, engaged in counterinsurgency efforts against JNIM and other militant groups
Malian Defense and Security Forces (MDSF)	The combined forces of Mali's military and police
Movement for Oneness and Jihad in West Africa (MUJAO)	A jihadist non-state armed group active in West Africa, later merged into JNIM
National Committee for the Salvation of the People (CNSP)	The military junta that performed a coup and led Mali in 2020–2021
Non-State Armed Group (NSAG)	A term describing armed organizations that use force to achieve their objectives, operate outside the direct control of a state, and are not formally integrated into regular armed forces or security institutions
Private Military Company (PMC)	Private enterprises providing military services, such as the Russian Wagner Group in Mali
Signals Intelligence (SIGINT)	Intelligence gathered from electronic signals, often used in counterinsurgency operations

Subject Matter Expert (SME)	Specialists providing critical insights in intelligence analysis or specific operational contexts
United Nations Multidimensional Integrated Stabilization Mission in Mali (MINUSMA)	A UN peacekeeping mission aimed at stabilizing Mali after the 2012 crisis, concluded in 2023

Chapter 1
Introduction

Over the past few decades, the Sahel region has become a hotbed of Islamist violence. Mali can be considered the original epicenter of this unfortunate development and a frontrunner in the effects of this increased instability (Congressional Research Service, 2023). The government's inability to effectively combat the insurgency in the northern and central regions of the country, led by various armed groups including Islamist militants, was a significant factor leading to the coup in Mali in August 2020 (Korotayev & Khokhlova, 2022). The military perceived the government's response to be inadequate and lacking strategic direction, exacerbated by the feeling among many military personnel that their sacrifices in combating insurgents were not being adequately recognized or addressed by the government (Korotayev & Khokhlova, 2022). Unsurprisingly, the coup leaders, known as the National Committee for the Salvation of the People (CNSP), justified their actions by citing the need to address, besides poor governance, the country's security challenges (Korotayev & Khokhlova, 2022).

Under international pressure, the CNSP agreed to a transitional process, which included the appointment of a civilian president and the formation of a transitional government tasked with organizing new elections and implementing reforms. However, tensions were high between the military and the civilian transitional government (Korotayev & Khokhlova, 2022), which led to a second coup in May 2021, when the military took full control over the government.

The second coup also signaled a pivot in Mali's geopolitical orientation. Frustrated by the deteriorating security situation and unconvinced that either France or the United Nations Multidimensional Integrated Stabilization Mission in Mali (MINUSMA for short) were able to help reverse the process, the Malian junta turned to Russia for assistance in countering the expanding insurgency (Ramani, 2020). In August 2022, the French armed forces left Mali, and on December 31 2023, based on the junta's request and thereafter mandated by the UN Security Council, MINUSMA closed. This left the Malian junta, assisted by the Russian mercenary organization Wagner, with the task of countering the Islamist violence (Ramani, 2020).

One of the most influential actors in spreading Islamist violence across the Sahel is Jama'at Nasr Al Islam Wal Muslimin (JNIM) (Hansen, 2019). This non-state armed group (NSAG) is the result of a merger in 2017 of four individual NSAGs that pledged allegiance to Al Qaeda, namely the Sahara Branch of Al-Qaeda in the Islamic Maghreb (AQIM), Ansar Dine, Al-Mourabitoun, and the Katiba Macina. Each of these Salafist extremist groups can be associated with one of the region's four most influential ethnic groups.

Thus, JNIM can be seen as a 'joint venture' of four distinct NSAGs that pursue diverse goals under the umbrella of Salafist extremism. Subsequently, its activities deliberately span the military, economic, and political domains, reflecting a multi-faceted approach to achieving its objectives in the Sahara-Sahel region. As a logical result, the organization engages in a comprehensive range of violent, criminal, political, administrative, and informational activities.

The 2017 merger signaled a rise in Islamist violence in the Sahara-Sahel region. To illustrate this fact, we obtained attack data from the well-known ACLED database.[1] Figure 1.1 below shows the evolution of JNIM attacks across the region

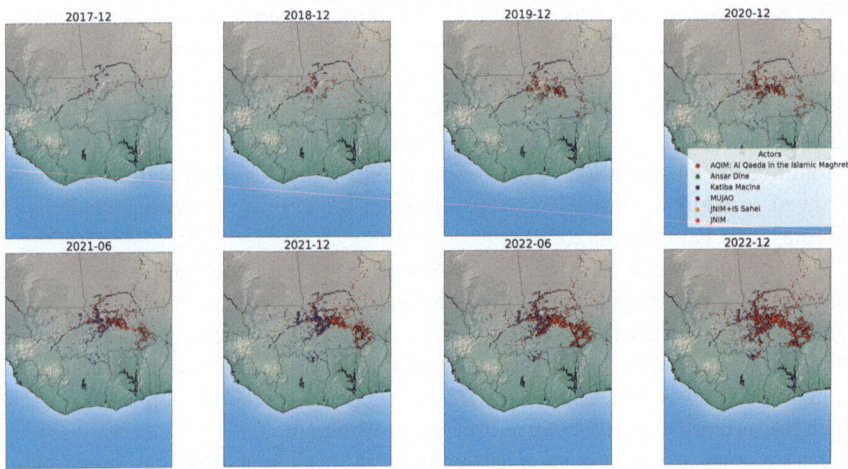

Fig. 1.1 Evolution of JNIM activity during the years 2017 (top left) to 2020 (top right), and half-yearly 2021 (bottom left) to 2022 (bottom right). Violent events attributed to JNIM are linked (red edge) if they occur within a timeframe of 30 days *and* within a 40 km radius of each other. The images, generated using the Basemap library in Python's Matplotlib, clearly indicate the increasing intensity and geographic dispersion of JNIM activities especially during the last couple of years, threatening the region's stability. Data and map sources are integrated from publicly available datasets and geospatial mapping resources

---

[1] accleddata.com

1 Introduction

during the 2015–2022 time period. Events are plotted as a geolocated point and are linked together if they occur within certain time-space constraints.[2]

The increasing insecurity in the Sahara-Sahel region is clearly visible. As a result, the region, faced with growing security problems, compounded by economic failures and widespread corruption, witnessed an increasing number of military coups, such as those in Mali (2020, 2021), Guinea (2021), Burkina Faso (2022), Niger (2023) and Gabon (2023). These coups and the subsequent international and internal instability created a power vacuum which Jihadist movements like JNIM have taken advantage of.

To provide some brief insights into the dynamics and the geographic expansion of these events, three basic variables are computed (see appendix D for more details). First, the *average* betweenness centrality (Brandes, 2001) of nodes in the evolving network, capturing the notion that an event leads to another event within certain time and space limits, i.e. encapsulating the *global* connectedness of initially dispersed conflict areas. Figure 1.2 shows an increasing trend, indicating the growing connectedness between events. Second, the *average* clustering coefficient

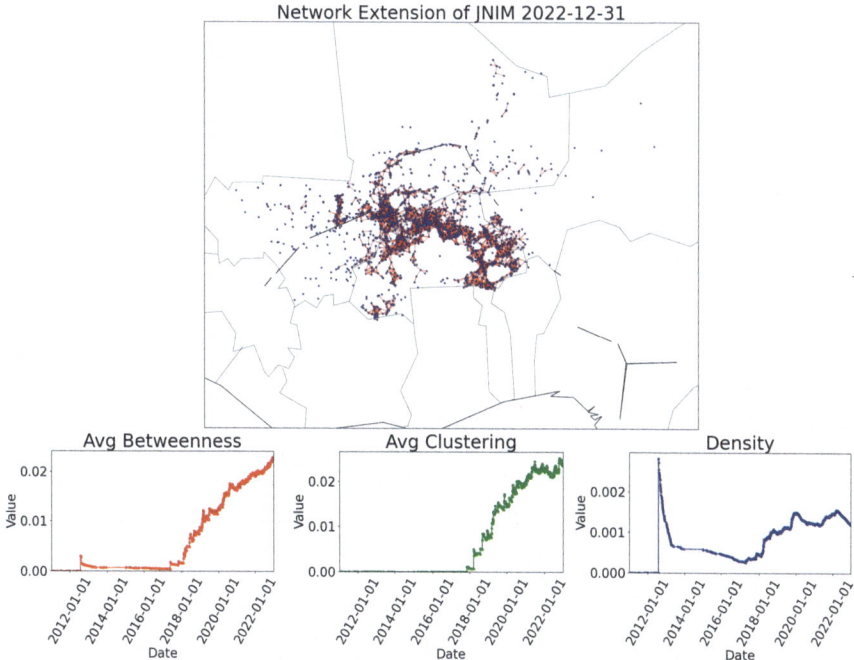

Fig. 1.2 Evolution of JNIM event network during the years 2015 to 2022 as measured by three spatio-temporal dimensions: average betweenness (left), average clustering (middle) and density (right). The figure was generated using Python's Matplotlib library, incorporating publicly available data sources

---

[2] During each year an event is plotted as a node and connected to another node if it occurred within 40 km and 180 days of that event.

(Opsahl & Panzarasa, 2009) that captures *local* connectedness between events also shows an increasing trend. Finally, we computed the density (Boguná et al., 2004) of the network, indicating the general amount of relatedness amongst insurgent events, which can also be seen to be increasing over time. These developments should worry decision makers. JNIM's observed geographic expansion into various regions, including Mali, Niger, Burkina Faso, and beyond, allowed the insurgency to overstretch counterinsurgency efforts and the required military forces to provide the preconditional basic security (Zimmerer, 2022). Its strategic and tactical maturity is evident in the systematic, coordinated expansion of its operations (as can be seen from Fig. 1.1), establishing operational bases, training camps and administrative structures in strategically important areas. The group interacts with other non-state armed groups in its area of operations, collaborating with Ansaroul Islam and competing with the Islamic State in the Greater Sahara (ISGS) (Pollicini, 2021). These interactions, despite ideological differences, underscore JNIM's influence and power in the Sahara-Sahel region, and in part explain the evolution of the metrics as shown in Fig. 1.2.

Numerous notable works have delved into JNIM, highlighting its characteristics and role in the Sahel (Hansen, 2019; Thurston, 2020; Zenn, 2022; Zimmerer, 2022). The foundations for data-driven, machine learning based analysis of terrorist groups were laid in prior studies of Hezbollah in 2008–2009 (Mannes et al., 2008; Mannes & Subrahmanian, 2009), Lashkar-e-Taiba (Subrahmanian et al., 2013), Indian Mujahideen (Subrahmanian et al., 2013), and Boko Haram (Subrahmanian et al., 2020). This book follows the methodological advances developed in those prior efforts and offers a powerful data-driven, model-based approach applied to the analysis of JNIM. By reducing data about a terrorist group into a spreadsheet and employing machine learning algorithms, the book aims to derive behavioral models, generate forecasts of JNIM's actions, and devise methods to shape counterinsurgency policies.

Our data spans a period of 12 years from January 2011 to December 2022 and includes data about 47 different type of attacks and a total 83 independent variables, or environmental variables, in all. Even though JNIM was officially established in March 2017, starting the dataset in 2011 ensures including information on the constituent organizations that later merged to form JNIM. By including data from this earlier period, we provide a more comprehensive view of relevant developments and operational patterns prior to JNIM's formal establishment.

1.1 Organization of this Book

The remainder of this book is structured as follows:

Chapter 2 offers an in-depth exploration of the origins of JNIM. After beginning with a concise overview of the Sahara-Sahel region, it proceeds to examine the intricate relationship between JNIM and Al Qaeda (AQ). The chapter delves into

1.1 Organization of this Book

the origins, motivations, and ideologies of JNIM, its operational geography, leadership, and the chronological evolution of its insurgency. It also discusses the political, military, and criminal dimensions of JNIM, essentially providing a concise history of the organization.

Chapter 3 introduces the Temporal Probabilistic (TP) rule formalism (Subrahmanian et al., 2013). This chapter, primarily for those interested in the technological aspects, explains how these rules were automatically derived from the JNIM dataset. Readers who are not interested in the technological basis of this book can safely skip this chapter.

Chapter 4 focuses on a specific category of attacks—abductions and the subsequent release of abductees—carried out by JNIM. JNIM carried out abductions in 89 of the 144 months included in our study. This chapter provides insights into the factors associated with abductions. The chapter identifies key factors, such as travel bans, asset freezes, and desertion by security forces. It presents complex rules that prove effective in explaining abductions and the release of abductees by JNIM.

Chapter 5 explores another class of attacks, specifically those targeting public sites. In 47 out of the 144 studied months, JNIM conducted attacks on public sites. It also discusses various factors linked to attacks on public sites in subsequent months.

Chapter 6 describes the targeting of security professionals or installations, a phenomenon occurring in over half of the studied months. Factors such as travel bans, asset freezes, executions by government security forces, and JNIM's claims of responsibility are examined in detail as predictors of future targeting of security forces and installations.

Chapter 7 focuses on the targeting of civilians for reasons not linked to their specific identity, yet another major instrument of terror used by JNIM (the targeting of civilians occurred in 77 out of the 144 studied months). This chapter focuses on identifying factors that are predictive of months where JNIM targets civilians for other reasons than their belief, political orientation or profession.

Chapter 8 concentrates on factors predicting three specific types of attacks: hit-and-run attacks, targeting of public transportation, and sabotage of facilities. The chapter introduces temporal probabilistic rules capable of anticipating these diverse attack scenarios.

Chapter 9 reflects on the implications of using these kinds of models in military decision-making and explores various aspects of predicting the behavior of non-state armed groups. Additionally, it aligns the used codebook[3] with doctrines for insurgency to enhance the analytical and predictive value of the presented model.

---

[3] The codebook contains a list of definitions of dependent and independent variables used in the data collection process. Please see Appendix B for a more detailed description.

1.2 How to Read this Book

For readers who are not inclined towards the technical aspects of the model used for deriving the results in this book, Chap. 3 can be comfortably skipped without missing any key insights into JNIM's behavior and activities. However, those with an interest in the technology will find the technical details presented in Chap. 3 interesting. While we do not delve into the mathematics underpinning our algorithms in this section, we provide clear explanations to convey the fundamental concepts behind these computations. For those seeking a deeper understanding, the technical references in these chapters offer further details. Table 1.1 below shows the number of attacks with respect to the different kinds of attacks captured in the data—this attack data was derived from the ACLED dataset.

Table 1.1 Different types of attacks carried out by JNIM, 2011–2022

Attacks	2011	2012	2013	2014	2015	2016	2017	2018	2019	2020	2021	2022
Abductions	0	0	0	0	3	1	1	6	12	11	10	12
Arson	0	1	0	0	1	2	1	8	11	12	11	11
Assassinations	0	0	0	0	2	2	0	8	11	9	10	9
Attempted Bombings	0	0	1	1	1	4	2	1	0	0	1	4
Casualties—Civilian	1	2	4	0	2	2	4	4	8	6	12	12
Casualties—Security Forces	0	2	3	0	3	1	2	4	2	3	2	9
Looting	0	0	0	0	0	0	0	4	4	3	11	10
Releases	0	1	0	1	0	0	1	0	4	8	12	12
Sexual Violence	0	0	0	0	0	0	0	0	1	0	6	4
Targ Civ for Beliefs	0	0	0	0	1	0	2	3	9	5	4	10
Targ Gov Officials	0	1	1	0	0	0	3	6	11	9	10	12
Targ Public Site	0	0	2	1	3	3	2	9	12	11	12	12
Targ Public Transport	0	0	0	0	0	0	0	1	4	4	3	3
Targ Security Installations	0	2	0	1	4	10	10	12	12	11	12	12

1.3 Summary Statistics about JNIM's Violent Activities

The study in this book derives temporal probabilistic (TP) rules[4] using data from the 2011–2022 time period. In this section, we will drill further down into the most prominent types of attacks, for which good predictive models were obtained. Figure 1.3, which outlines the number of occurrences of abductions, hit-and-run attacks, and assaults on public places between 2011 and 2022, visualizes a noticeable shift. Up until 2017, incidences of these attacks remained limited. However, post-2017, there was a substantial surge in both variants of attacks and abductions, indicating a significant change in JNIM's operational dynamics. Chap. 4 will further investigate the occurrence of abductions. Chap. 5 will discuss targeting of and attacks on public places and Chap. 8 will explore the occurrences of hit-and-run attacks in greater detail.

Figure 1.4 details the occurrences of attacks targeting civilians without specific reasons, targeting public places, targeting security professionals, and targeting security installations on a monthly basis. In this figure, a similar trend emerges. The years leading up to 2017 witnessed a relatively subdued frequency of these types of targeting by JNIM. Post-2017, however, all four categories experienced a notable escalation. Of particular interest is the sharp increase in the targeting of security installations, with targeting of security professionals closely following suit. Notably,

Fig. 1.3 Number of months (2011–2022) when JNIM carried out abductions, hit and run attacks and attacks on public sites

---

[4]A detailed description of a TP-rule will be provided in Chap. 3

Fig. 1.4 Number of months (2011–2022) when JNIM targeted public sites, security professionals, security installations or civilians

the targeting of civilians without specific reasons linked to their profession, belief, or political orientation exhibited the most pronounced and sustained upward trajectory, suggesting a shift in JNIM's strategic focus during this period.

In general, instances of violent behavior connected to JNIM experienced a significant increase around or after 2017. There are likely multiple factors contributing to this trend. Firstly, during this period, several groups united under the banner of JNIM, potentially streamlining the documentation of their activities by news agencies. Additionally, this consolidation may have marked the initiation of the subsequent phase that AQ initiated in the Sahel. Further insights that may help address these questions will be discussed in Chap. 2.

1.4 Summary of Significant Temporal Probabilistic (TP) Rules

Though we derived thousands of valid TP-rules, we selected the 37 most interesting TP-rules that help explain the conditions[5] under which ten types of actions are taken by JNIM. These actions include:

- Abduction
- The release of abductees

---

[5] For a detailed overview off all conditions identified, please see Appendix C

1.4 Summary of Significant Temporal Probabilistic (TP) Rules

- Targeting of public sites
- Attacks on public sites
- Targeting of security professionals
- Targeting of security installations
- Targeting of civilians
- Hit and run attacks
- Sabotages
- Targeting of public transportation facilities

In our studies, a total of 17 conditions are linked to the aforementioned actions of JNIM. Of these, the most significant factors are the following: travel bans placed on the state where JNIM operates, assets of JNIM being frozen by foreign states or international institutions, the government raiding facilities or locations of JNIM and the national government of the state where JNIM operates receiving foreign military aid. We discuss each of these below.

1.4.1 Travel Bans Placed on Mali, Niger or Burkina Faso

Over the past decade, numerous foreign states, particularly neighboring ones, have imposed travel bans on Mali, Niger, or Burkina Faso. Our findings reveal that these travel bans have a significant impact on almost every action taken by JNIM discussed in this book. To illustrate, travel bans imposed on Mali are negatively linked with the occurrence of abductions several months later. Meaning, when a travel ban is imposed, different kind of attacks could occur several months later. Additionally, instances of attacks on public sites appear to be more likely when travel bans were placed on Mali, Burkina Faso, or Niger in the preceding months. These results suggest that implementing travel bans on states where JNIM operates may have counterproductive effects in terms of reducing various forms of violent activities, such as abductions and attacks on public sites.

1.4.2 Freezing Assets of (Members of) JNIM by Foreign States or International Institutions

In accordance with resolutions from the UN Security Council, several foreign states have taken steps over the past few years to freeze the assets of individual members of JNIM. Our findings indicate a negative linkage between the freezing of JNIM members' assets and their engagement in activities such as targeting civilians, abductions, hit-and-run attacks, and assaults on public sites. Furthermore, when assets are not subjected to freezes, a positive effect is observed in terms of the targeting of security professionals and/or installations.

1.4.3 The National Government of the State where JNIM Operates Orders Executions of Declared Offenders

In their attempts to maintain order over the past decade, security forces in Mali, Niger, and Burkina Faso have executed or killed civilians (Amnesty International, 2020). These executions often occurred without a clear legal process or verdict. Simultaneously, these actions significantly impact the population and influence the behavior of groups like JNIM. In this study, we have identified a negative link between the execution of 'declared' offenders and the targeting of security installations or professionals. In other words, we observed that the execution of declared offenders correlates with a rise in the targeting of security personnel and structures.

1.4.4 The National Government where JNIM Operates Receives Foreign Military Aid

As mentioned before, Mali has been receiving foreign military aid through MINUSMA, as well as through a number of bilateral agreements with other countries over the past decade. Foreign military aid, both in terms of physical support and financial assistance, has played a significant role in shaping JNIM's campaign. The group actively opposes foreign aid in its campaigns and has also targeted UN or MINUSMA troops—perhaps because they support the Malian government. Initial observations suggest that foreign military aid is a crucial factor influencing JNIM's actions. For instance, our findings indicate that hit-and-run attacks become more frequent as the national government receives increased foreign military aid. Additionally, civilians are targeted more often when foreign military aid is provided.

1.5 Efficacy of our Predictive Modeling & the NTEWS System

The Temporal Probabilistic(or TP) rules referenced above help explain the conditions under which JNIM carries out different types of attacks. However, to make good predictions, we trained 6 well-known machine learning models (Random Forest, KNN, SVM, Multinomial Naive Bayes, Gaussian Naive Bayes, Logistic Regression) on the January 2011–December 2022 timeframe. In addition, we developed a "late fusion" model that combines the results predicted by these six basic machine learning models. This led to a total of 42 trained models, one each to predict attacks that will occur sometime in the next month, sometime in the next 2 months, ..., sometime in the next 6 months. It is important to note that during training, we learned which of the machine learning models is best for a given attack and time period t, ranging from 1–6 months. We have developed a system called

1.5 Efficacy of our Predictive Modeling & the NTEWS System

NTEWS (Northwestern Terror Early Warning System) and made the monthly reports generated by NTEWS publicly available on October 17 2024. NTEWS provides predictions of attacks by six terrorist groups: Abu Sayyaf, Al Shabaab, Boko Haram, Indian Mujahideen, JNIM, and Lashkar-e-Taiba. If you are interested in signing up for NTEWS reports, you can do so at https://docs.google.com/forms/d/1 7fiO1AX7aE5vm71CqvJQPgn7KwUZaDOwv-enJGXvL7Q/ edit?ts=6740a3a6&pli=1. More information on NTEWS can be seen at https://sites.northwestern.edu/nsail/projects/ntews/.

We then tested the best machine learning model in NTEWS for each pair (A, T) where A is an attack type[6] and T is a time period.

Table 1.2 below shows the efficacy of our learned predictive models when they were used to predict attacks by JNIM from January 2023 to December 2023 *one month into the future*. This table shows 4 metrics. Precision refers to the probability that a prediction that an attack will occur is correct. For instance, if we look at the precision of 0.74 for assassinations in Table 1.2, it means that 74% of the times when our system predicted an assassination were correct, i.e. an assassination did occur sometime in the next month after the prediction was made. Recall, on the other hand, is the probability that a real attack was correctly predicted by the NTEWS system. For instance, if we look at the recall of 0.81 for abduction in Table 1.2, we see that of all the months in which an abduction took place in the next month, NTEWS successfully predicted 81% of them. AUC (which stands for Area Under a Receiver Operating Characteristic curve) and F1 score are more complex measures. F1 Score is a combination of precision and recall. In all four cases, scores lie between 0 and 1 and a score closer to 1 is more desirable.

The reader can readily see that NTEWS performs very well on certain types of attacks—and less well on others. Predictive performance on certain types of attacks may be poor because of several possible reasons. One possibility is that the number of occurrences of that type of attack might be very small. Another possibility is that the data might be highly imbalanced with either the number of months in which the attack occurred by vastly bigger (or vastly smaller) than the number of months in which the attack didn't occur. Simply put, machine learning algorithms have not yet achieved high predictive performance in these types of cases.

We use the F1 score (with a 75% cutoff) as the metric to use to decide whether NTEWS should be used to predict a certain kind of attack or not, NTEWS can make reliable predictions of whether an attack might occur sometime in the next month for 18 types of attacks.

Table 1.3 below shows a trimmed version of Table 1.2 that focuses only on the types of attacks that can be predicted to occur sometime in the next *two* months with an F1-score exceeding 75%.

In fact, Table 1.3 shows that we can predict whether 22 types of attacks will occur sometime in the next 2 months. The fact that this reflects an increase in the number of attack types compared to the 18 we could predict well sometime in the

---

[6] Please note that "Release of Abductees" is included in the list of "attacks" even though, of course, this is a desirable action from a counterterrorism perspective by JNIM.

Table 1.2 Predictive performance of NTEWS on whether a given type of attack will occur sometime in the next month

	Month 1			
	AUC	Precision	Recall	F1
Abduction	0.91	1.00	0.81	0.90
Release of Abductees	0.73	0.67	0.60	0.63
Assassination	0.85	0.74	1.00	0.85
Sexual Violence	0.57	0.25	0.25	0.25
Armed Clashes with Group Casualties	0.86	0.84	0.94	0.89
Armed Clashes with Security Forces Casualties	0.73	0.80	0.50	0.62
Armed Clashes with Unspecified Casualties	0.82	0.72	0.93	0.81
Arson	0.90	0.92	0.86	0.89
Unspecified Attack	0.70	0.80	0.44	0.57
Attempted Attack	0.80	0.75	0.75	0.75
Civilian Casualties	0.83	0.85	0.79	0.81
Attack on Government	1.00	1.00	1.00	1.00
Hit & Run Attack on Security Forces	0.90	0.89	0.94	0.91
Attack on Worksites	0.73	0.50	0.50	0.50
Attacks on Public Sites	0.83	0.69	0.92	0.79
Attacks on an Education Facility	0.50	0.00	0.00	0.00
Attempted Bombing	0.63	1.00	0.25	0.40
Bombing	0.94	0.93	0.93	0.93
Suicide Bombing	0.44	0.05	0.50	0.10
Looting	0.79	0.75	0.67	0.71
Robbery	0.92	0.80	1.00	0.89
Sabotage	1.00	1.00	1.00	1.00
Seizure of Security Force Installations	0.50	0.00	0.00	0.00
Seizure of Territory	0.50	0.00	0.00	0.00
Targeting Civilians for their Belief	0.72	0.46	0.75	0.57
Targeting Civilians for Other Reasons	0.84	0.88	0.82	0.85
Targeting Civilians in Mass Casualty Attacks	0.75	1.00	0.50	0.67
Targeting Civilians for their Political Orientation	0.70	0.71	0.50	0.59
Targeting Government Officials	0.89	0.76	1.00	0.87
Targeting Security Forces	0.97	1.00	0.95	0.97
Targeting Civilians for their Profession	0.75	1.00	0.50	0.67
Targeting Teachers	0.67	1.00	0.33	0.50
Targeting Government Buildings	0.67	0.60	0.43	0.50
Targeting Security Installations	0.88	0.94	0.84	0.89
Targeting Work Sites	0.67	1.00	0.33	0.50
Targeting Public Sites	0.84	0.87	0.81	0.84
Targeting Public Transport	0.76	0.60	0.60	0.60
Targeting Other Structures	0.75	1.00	0.50	0.67
Targeting Symbolic Sites	0.82	0.50	0.75	0.60

1.5 Efficacy of our Predictive Modeling & the NTEWS System

Table 1.3 Predictive performance of NTEWS on whether a given type of attack will occur sometime in the next 2 months—only attacks that can be predicted with an F1 score of 75% or more are shown

	Month 2			
	AUC	Precision	Recall	F1
Abduction	0.84	0.89	0.84	0.86
Release of Abductees	0.93	0.92	0.92	0.92
Assassination	0.85	0.93	0.78	0.85
Sexual Violence	0.80	1.00	0.60	0.75
Armed Clashes with Group Casualties	0.83	0.87	0.95	0.91
Armed Clashes with Unspecified Casualties	0.84	0.81	0.87	0.84
Arson	0.87	0.93	0.81	0.87
Attempted Attack	0.94	0.88	1.00	0.94
Hit & Run Attack on Security Forces	0.91	0.94	0.89	0.91
Attacks on Public Sites	0.81	0.74	0.93	0.82
Bombing	0.91	1.00	0.82	0.90
Looting	0.83	0.73	0.80	0.76
Robbery	0.93	0.92	0.92	0.92
Sabotage	0.93	1.00	0.86	0.92
Targeting Civilians for their Belief	0.86	0.83	0.83	0.83
Targeting Civilians for Other Reasons	0.93	1.00	0.85	0.92
Targeting Government Officials	0.97	1.00	0.93	0.97
Targeting Security Forces	0.88	0.94	0.85	0.89
Targeting Civilians for their Profession	0.89	0.62	1.00	0.76
Targeting Security Installations	0.90	0.95	0.91	0.93
Targeting Public Sites	0.91	0.94	0.89	0.91
Targeting Other Structures	0.80	1.00	0.60	0.75

next month because these predictions exhibit greater temporal uncertainty. As an example of why this coarser grained temporal granularity can sometimes enable better prediction, consider sexual violence in Tables 1.2 and 1.3. The F1-score for predicting sexual violence sometime in the next month is only 0.25 (Table 1.2) but when we allow the prediction of sexual violence to be more temporally uncertain, i.e. predict if it will or will not occur sometime in the next 2 months, then the F1-score goes up to 0.75 (Table 1.3).

Table 1.4 shows that we can predict whether 29 types of attacks will occur sometime in the next 3 months. Again, this prediction has greater temporal uncertainty, explaining why more attacks can be predicted well within this timeframe.

Table 1.5 shows that we can predict whether 31 types of attacks will occur sometime in the next 4 months. Again, this prediction has greater temporal uncertainty, explaining why more attacks can be predicted well within this timeframe.

Table 1.6 shows that we can predict whether 31 types of attacks will occur sometime in the next 5 months.

Table 1.4 Predictive performance of NTEWS on whether a given type of attack will occur sometime in the next 3 Months—only attacks that can be predicted with an F1 score of 75% or more are shown

	Month 3			
	AUC	Precision	Recall	F1
Abduction	0.81	0.86	0.90	0.88
Release of Abductees	0.85	1.00	0.69	0.82
Assassination	0.86	0.90	0.90	0.90
Sexual Violence	0.86	1.00	0.71	0.83
Armed Clashes with Group Casualties	0.89	0.95	0.91	0.93
Armed Clashes with Unspecified Casualties	0.88	0.93	0.82	0.87
Arson	0.77	0.78	0.82	0.80
Unspecified Attack	0.87	0.85	0.85	0.85
Attempted Attack	0.84	0.92	0.75	0.83
Civilian Casualties	0.79	0.85	0.85	0.85
Attack on Government	0.80	1.00	0.60	0.75
Hit & Run Attack on Security Forces	0.97	1.00	0.95	0.97
Attacks on Public Sites	0.84	1.00	0.69	0.81
Attacks on an Education Facility	0.98	0.75	1.00	0.86
Attempted Bombing	0.87	0.88	0.78	0.82
Bombing	0.92	1.00	0.83	0.91
Looting	0.85	0.90	0.75	0.82
Robbery	0.94	0.88	1.00	0.93
Sabotage	0.87	0.88	0.78	0.82
Targeting Civilians for their Belief	0.85	0.76	0.93	0.84
Targeting Civilians for Other Reasons	0.90	0.95	0.91	0.93
Targeting Government Officials	0.91	1.00	0.81	0.90
Targeting Security Forces	0.88	0.91	0.95	0.93
Targeting Teachers	0.88	0.77	0.91	0.83
Targeting Government Buildings	0.79	0.60	1.00	0.75
Targeting Security Installations	0.88	0.95	0.86	0.90
Targeting Work Sites	0.81	0.69	0.82	0.75
Targeting Public Sites	0.88	0.95	0.86	0.90
Targeting Public Transport	0.85	0.67	0.89	0.76

1.5 Efficacy of our Predictive Modeling & the NTEWS System

Table 1.5 Predictive performance of NTEWS on whether a given type of attack will occur sometime in the next 4 months—only attacks that can be predicted with an F1 score of 75% or more are shown

	Month 4			
	AUC	Precision	Recall	F1
Abduction	0.90	0.95	0.90	0.93
Release of Abductees	0.86	1.00	0.71	0.83
Assassination	0.70	0.81	0.95	0.88
Sexual Violence	0.85	0.86	0.75	0.80
Armed Clashes with Group Casualties	0.86	0.92	1.00	0.96
Armed Clashes with Security Forces Casualties	0.78	0.86	0.71	0.77
Armed Clashes with Unspecified Casualties	0.87	0.88	0.88	0.88
Arson	0.85	0.93	0.78	0.85
Attempted Attack	0.87	0.88	0.88	0.88
Civilian Casualties	0.64	0.78	0.95	0.86
Hit & Run Attack on Security Forces	0.97	1.00	0.95	0.97
Attack on Worksites	0.83	1.00	0.67	0.80
Attacks on Public Sites	0.76	0.86	0.67	0.75
Attacks on an Education Facility	1.00	1.00	1.00	1.00
Bombing	0.85	0.94	0.79	0.86
Looting	0.91	0.85	0.92	0.88
Robbery	0.94	0.88	1.00	0.93
Sabotage	0.89	1.00	0.78	0.88
Targeting Civilians for their Belief	0.80	0.78	0.88	0.82
Targeting Civilians for Other Reasons	0.86	0.92	1.00	0.96
Targeting Civilians for their Political Orientation	0.80	0.83	0.83	0.83
Targeting Government Officials	0.91	1.00	0.82	0.90
Targeting Security Forces	0.95	1.00	0.91	0.95
Targeting Civilians for their Profession	0.81	0.80	0.73	0.76
Targeting Teachers	0.88	0.79	0.92	0.85
Targeting Government Buildings	0.91	0.82	1.00	0.90
Targeting Security Installations	0.85	0.92	0.96	0.94
Targeting Work Sites	0.88	0.79	0.92	0.85
Targeting Public Sites	0.73	0.85	0.96	0.90
Targeting Public Transport	0.83	0.73	0.80	0.76
Targeting Other Structures	0.83	1.00	0.67	0.80

Table 1.6 Predictive performance of NTEWS on whether a given type of attack will occur sometime in the next 5 months—only attacks that can be predicted with an F1 score of 75% or more are shown

	Month 5			
	AUC	Precision	Recall	F1
Abduction	0.77	0.88	0.91	0.89
Assassination	0.81	0.88	1.00	0.94
Sexual Violence	0.81	1.00	0.63	0.77
Armed Clashes with Group Casualties	0.83	0.93	1.00	0.96
Armed Clashes with Security Forces Casualties	0.81	0.82	0.95	0.88
Armed Clashes with Unspecified Casualties	0.88	0.94	0.83	0.88
Arson	0.88	0.94	0.84	0.89
Unspecified Attack	0.83	0.83	0.88	0.86
Attempted Attack	0.84	0.88	0.83	0.86
Civilian Casualties	0.58	0.83	1.00	0.91
Hit & Run Attack on Security Forces	0.95	1.00	0.90	0.95
Attacks on Public Sites	0.85	0.82	1.00	0.90
Attacks on an Education Facility	1.00	1.00	1.00	1.00
Bombing	0.90	1.00	0.80	0.89
Looting	0.92	0.81	1.00	0.90
Robbery	1.00	1.00	1.00	1.00
Sabotage	0.90	0.82	0.90	0.86
Targeting Civilians for their Belief	0.88	0.93	0.82	0.87
Targeting Civilians for Other Reasons	0.92	0.96	1.00	0.98
Targeting Civilians for their Political Orientation	0.59	0.69	1.00	0.82
Targeting Government Officials	0.91	1.00	0.82	0.90
Targeting Security Forces	0.75	0.85	1.00	0.92
Targeting Civilians for their Profession	0.95	0.86	1.00	0.92
Targeting Teachers	0.92	0.81	1.00	0.90
Targeting Government Buildings	0.94	0.93	0.93	0.93
Targeting Security Installations	0.81	0.91	0.87	0.89
Targeting Work Sites	0.92	1.00	0.85	0.92
Targeting Public Sites	0.69	0.85	0.96	0.90
Targeting Public Transport	0.93	0.91	0.91	0.91
Targeting Other Structures	0.87	0.67	0.86	0.75
Targeting Symbolic Sites	0.84	0.79	0.85	0.81

Table 1.7 shows that we can predict whether 35 types of attacks will occur sometime in the next 6 months. Again, this prediction has greater temporal uncertainty, explaining why more attacks can be predicted well within this timeframe.

1.5 Efficacy of our Predictive Modeling & the NTEWS System

Table 1.7 Predictive performance of NTEWS on whether a given type of attack will occur sometime in the next 6 months—only attacks that can be predicted with an F1 score of 75% or more are shown

	Month 6			
	AUC	Precision	Recall	F1
Abduction	0.86	0.95	0.87	0.91
Release of Abductees	0.83	0.92	0.73	0.81
Assassination	0.79	0.88	1.00	0.94
Sexual Violence	0.81	1.00	0.63	0.77
Armed Clashes with Group Casualties	0.88	0.96	0.96	0.96
Armed Clashes with Security Forces Casualties	0.73	0.79	0.95	0.86
Armed Clashes with Unspecified Casualties	0.88	0.94	0.83	0.88
Arson	0.81	0.85	0.89	0.87
Unspecified Attack	0.81	0.87	0.76	0.81
Attempted Attack	0.85	0.93	0.78	0.85
Civilian Casualties	0.88	0.96	1.00	0.98
Hit & Run Attack on Security Forces	0.97	1.00	0.95	0.97
Attack on Worksites	0.93	1.00	0.86	0.92
Attacks on Public Sites	0.82	0.93	0.74	0.82
Attacks on an Education Facility	1.00	1.00	1.00	1.00
Attempted Bombing	0.80	0.77	0.77	0.77
Bombing	0.85	0.90	0.90	0.90
Looting	0.94	0.87	1.00	0.93
Robbery	0.97	1.00	0.93	0.97
Sabotage	0.90	1.00	0.80	0.89
Targeting Civilians for their Belief	0.86	0.88	0.88	0.88
Targeting Civilians for Other Reasons	0.70	0.89	1.00	0.94
Targeting Civilians in Mass Casualty Attacks	0.86	0.67	0.80	0.73
Targeting Civilians for their Political Orientation	0.59	0.74	0.95	0.83
Targeting Government Officials	0.91	1.00	0.82	0.90
Targeting Security Forces	0.91	0.95	0.95	0.95
Targeting Civilians for their Profession	0.91	0.81	1.00	0.90
Targeting Teachers	0.97	0.93	1.00	0.97
Targeting Government Buildings	0.90	0.88	0.93	0.90
Targeting Security Installations	0.81	0.91	0.91	0.91
Targeting Work Sites	0.97	0.93	1.00	0.96
Targeting Public Sites	0.83	0.92	1.00	0.96
Targeting Public Transport	0.94	0.86	1.00	0.92
Targeting Other Structures	0.91	0.86	0.86	0.86
Targeting Symbolic Sites	0.86	0.92	0.79	0.85

1.6 Implications for Military Decision Making

Using data-driven models in military decision making is becoming more prevalent and important. The huge amount of relevant data available during many contemporary military operations cannot be analyzed by humans alone. In addition to machine learning models that make predictions such as those described in Sect. 1.5, models that help to understand the behavior of adversaries, such as the temporal-probabilistic rules presented in this book, are extremely relevant. A combination of human-machine teaming in military operations and intelligence analysis is needed to stay one step ahead of our adversaries. In this book, the (JNIM specific) case will be made that to make optimal use of TP-rules that encapsulate NSAG (Non-State Armed Group) behavior military decision makers need to:

- Integrate TP-rules into the behavioral analysis of NSAGs.
- Allocate subject matter experts to coding teams that develop NSAG specific codebooks.
- Develop an iterative framework to update variable selection and codebook development during a military campaign.
- Educate and train military professionals and subject matter experts in the use of data-driven tools.

One note of caution is in order: any data-driven, machine learning based insights are correlative, not causal. The insights derived in this book must be complemented with appropriate expertise and insights that commanders in this field have learned over the years.

1.7 Conclusion

In the turbulent landscape of the Sahel, JNIM has distinguished itself as one of the most disruptive non-state armed groups, leaving a trail of deadly attacks that occurred almost weekly throughout 2022–2023. This book presents 12 years of monthly data from January 2011 to December 2022 and highlights the use of data science in counterinsurgency, leveraging computational models to predict and understand the diverse range of attacks orchestrated by JNIM. Furthermore, it explores the practical implementation of these models in informing the military decision-making processes. The discussion within this chapter focuses on key factors linked to JNIM's patterns of attacks. These factors encompass travel bans imposed on the state where JNIM operates, the freezing of JNIM's assets by foreign states or international institutions, government-led raids on JNIM facilities or locations, and the national government in the state where JNIM operates receiving foreign military aid.

The next chapters of this book will provide further insight into these factors while also explaining the model and TP-rules in more detail. The final chapter of this study is dedicated to a thorough examination of the implications associated with deploying such models in military decision-making processes. Through a careful evaluation of both strengths and weaknesses, the chapter aims to contribute valuable insights and concludes by presenting suggestions for future research.

References

Amnesty International. (2020). Human rights violations by security forces in the Sahel. In www.amnesty.org (AFR 37/2318/2020). https://www.amnesty.org/fr/wp-content/uploads/2023/06/AFR3723182020ENGLISH.pdf

Boguná, M., Pastor-Satorras, R., Díaz-Guilera, A., & Arenas, A. (2004). Models of social networks based on social distance attachment. *Physical Review E—Statistical, Nonlinear, and Soft Matter Physics, 70*(5), 056122.

Brandes, U. (2001). A faster algorithm for betweenness centrality. *Journal of Mathematical Sociology, 25*(2), 163–177.

Congressional Research Service. (2023). *Instability in the Sahel region: Mali as an epicenter*. Congressional Research Service.

Hansen, S. J. (2019). *Horn, Sahel and rift fault-lines of the African Jihad*. Hurst & Company.

Korotayev, A., & Khokhlova, A. (2022). Revolutionary events in Mali, 2020–2021. In *New wave of revolutions in the MENA region: A comparative perspective* (pp. 191–218). Springer International Publishing.

Mali: L'attaque meurtrière contre la Minusma revendiquée. (2019, January 21). RFI. https://www.rfi.fr/fr/afrique/20190121-mali-attaque-meurtriere-contre-onu-revendiquee-aqmi

Mannes, A., & Subrahmanian, V. S. (2009). Calculated terror. *Foreign Policy*. https://foreignpolicy.com/2009/12/15/calculated-terror/

Mannes, A., Michael, M., Pate, A., Sliva, A., Subrahmanian, V. S., & Wilkenfeld, J. (2008). Stochastic opponent modeling agents: A case study with Hezbollah. In *Social computing, behavioral modeling, and prediction* (pp. 37–45). Springer US.

Opsahl, T., & Panzarasa, P. (2009). Clustering in weighted networks. *Social Networks, 31*(2), 155–163.

Pollicini, L. (2021). A case of violent corruption: JNIM's insurgency in Mali (2017-2019). *Small Wars & Insurgencies, 32*(7), 1092–1116.

Ramani, S. (2020) Why Russia is a geopolitical winner in Mali's coup, Foreign Policy Research Institute, https://www.fpri.org/article/2020/09/why-russia-is-a-geopolitical-winner-in-malis-coup/

Subrahmanian, V. S., Mannes, A., Roul, A., & Raghavan, R. K. (2013). *Indian Mujahideen: Computational analysis and public policy*. Springer.

Subrahmanian, V. S., Pulice, C., Brown, J. F., & Bonen-Clark, J. (2020). *A machine learning based model of Boko Haram*. Springer Nature.

Thurston, A. (2020). *Jihadists of North Africa and the Sahel: Local politics and rebel groups*. Cambridge University Press.

Zenn, J. (2022). Rebel rivals: Reinterrogating the movement for unity and Jihad in West Africa and the roots of Al-Qaeda-Islamic state infighting in the Sahel. *The Journal of North African Studies, 27*(6), 1277–1301.

Zimmerer, M. (2022). Terror in West Africa: A threat assessment of the new Al Qaeda affiliate in Mali. *Critical Studies on Terrorism, 12*(3), 491–511.

Chapter 2
History of Jama'at Nasr al-Islam wal Muslimin (JNIM)

2.1 Introduction

Identifying the nature of a non-state armed group (NSAG) in the context of an ongoing insurgency is difficult. This difficulty is more pronounced when the conflict involves a blurring of the distinctions between war, organized crime, and large-scale violations of human rights. In particular, insurgencies often consist of multiple networks of multiple NSAGs, some of which have different roles in different networks (Bunker, 2005). Thus, it is hard to define such an actor with different roles in various networks and as a result, insurgent groups are often mislabeled as either a terrorist organization, or a criminal network.

This is also the case with the subject of this book, *Jama'a Nusrat ul-Islam wa al-Muslimin'* (JNIM), which translates into Group to Support Islam and Muslims (GSIM). JNIM is the main NSAG responsible for destabilizing the Sahara-Sahel region and was established in 2017 when four powerful NSAGs joined forces (ACLED, 2023). Several governments and international institutions have labeled JNIM as a terrorist organization, implying that it has narrow political goals (De ministers van Buitenlandse Zaken, van Defensie, voor Buitenlandse Handel & Ontwikkelingssamenwerking en van Veiligheid & Justitie, 2018; United Nations Security Council, 2019). In addition, some also argue that the regional AQ affiliates' ideological and political rhetoric merely serves as a cover for profitable criminal activities (Reitano et al., 2017; Harmon, 2014). This view is supported by academic literature and case studies, in which it is recognized that greed can be a cause for an insurgency and that leaders of violent non-state actors can be affected by opportunities for personal enrichment through illegal means (Jones, 2017). Thus, JNIM is

---

This chapter is in part derived from Provoost, Marnix. "*Symbiosis in the Sahara and Sahel: Why and How Does Jama'a Nusrat Ul-Islam Wa Al-Muslimin' (JNIM) Combine Insurgent, Terrorist and Criminal Activities and What Does This Reveal About the Movement's True Nature?*", MA Thesis, Netherlands Defence Academy, Breda, 2019.

often considered a terrorist organization that also uses Islamic rhetoric as a guise for personal enrichment through criminal activities.

However, one should focus on the operational level to qualitatively determine the nature of a non-state armed group (NSAG). At this level, it is possible to see the essence of a group and distinguish between the use of core and peripheral activities between objectives and modalities (Mulaj, 2010). The operational level of analysis enables distinguishing ends, means, and ways and thus reveals a group's strategy. In turn, this strategy reveals the group's nature. Academics and analysts who have more recently focused on this level of analysis, have come to a more nuanced appreciation of JNIMs engagement in organized crime (Provoost, 2019; Beevor, 2022; Nsaibia et al., 2023).

The academic literature provides several useful angles to analyze the organization's activities and assess JNIM's nature. One is to focus on the objectives, strategy, tactics, and leaders' agendas within the context of the environment in which JNIM operates (Mulaj, 2010). Another is to categorize JNIM based on factors such as motivation, stance towards change, and degree of organization (Bailes & Nord, 2010). Both angles are used in this chapter to create a rough outline of JNIM and its area of operations, thus providing a context for the subsequent quantitative data analysis of its activities.

Finally, due to JNIM's strong ties to Al-Qaeda in the Islamic Maghreb (AQIM) and the central AQ organization, it is essential to look at AQ's ideological views on war, declared goals, strategy and operational guidelines. JNIM's activities can then be assessed within the ideological, doctrinal and operational context from which they get additional purpose and meaning. Assessing JNIM within this context of AQ reveals a roadmap and thus has much analytical and to some degree even predictive value.

This context is provided by the translated writings and audio & video messages of the late Dr. Ayman al-Zawahiri and the late 'Abd Al-Aziz Al-Muqrin. The former was Bin-Laden's top lieutenant and fulfilled AQ's leadership position from 2011 until his death in 2022. He is seen as AQ's strategist who held significant sway over the group's operations (Baldauf, 2001). The latter stood out within AQ for the systematic and analytical approach via which he thought about warfare (Cigar, 2009, p.4). His work, *Dawrat al-tanfidh wa-harb al'asabat* [translated into *A Practical Course for Guerrilla Warfare*] is considered to codify a general consensus on doctrine within the then Al-Qaeda in the Arabian Peninsula (AQAP) leadership (Cigar, 2009, pp.12–13). Although al-Muqrin's significance diminished after he was killed by Saudi security forces in June 2004, his work has obviously found its way throughout AQ's central organization and subsections given its apparent application in practice. Perhaps the most striking strategic example is al-Muqrin's advocacy for a decentralized approach to the global jihad, emphasizing the importance of localized insurgencies rather than ones exported by the central organization.

Another useful source to provide context for JNIM are the so-called Timbuktu papers (Guidère, 2014). These papers are actually a cache of documents that belonged to AQIM, which were discovered in 2013 in Timbuktu. The documents contained administrative records, internal communications, and ideological manifestos, thus offering a rare glimpse into AQIM's ideology, strategy, and inner workings in the Sahara-Sahel region. The documents provide invaluable insights and context regarding JNIM and have been analyzed thoroughly (Guidère, 2014; Forbes, 2018).

2.2 The Sahara-Sahel Region

The Sahara-Sahel region is a vast, arid, and sparsely populated area that bridges the Mediterranean and Sub-Sahara parts of the African continent. Life in the region is challenging, yet it is home to nomadic, semi-sedentary, and sedentary ethnicities. The interaction between four of these groups, namely the Tuareg, Arabs, Fulani and Songhai, influences the Sahara-Sahel region's security situation. Although different in ethnicity, the four groups share several cultural similarities (Harmon, 2014). All four groups have an Islamic heritage, encompassing moderate and radical forms of Islam. Furthermore, all groups are internally divided by class, occupational specialization, and (perception of) race, as well as identification of a noble, free, and servile status. Finally, the effects of colonial rule and decolonization still resonate in the interaction between these four ethnic groups. All of the above factors translate into the root causes of the instability in the Sahara-Sahel region and have produced a multitude of armed groups from different backgrounds and with different motives.

The Sahara-Sahel region's economy is historically based on intra-regional connectivity and heavily influenced by the trans-Saharan trade between the Maghreb and the Niger valley. The region's economic processes have been disrupted by arbitrary borders, corrupt officials, poor infrastructure, and ethnic rivalries. As a result, large parts of the regional economy went underground. This has caused a subjective perception of relative degrees of illegality and morality, as well as ethnic and political conflict over access to the underground economy (Harmon, 2014). The ancient trade routes across the poorly controlled territory have always appealed to smugglers and robbers alike, and having control over these routes has always been a profitable source of income for any actor. Most of the criminal armed groups that seek control over the trade routes are rooted in the networks of different Sahrawi and Arab communities that can rely on family and tribal ties throughout the whole region (Lacher, 2012). This enables these networks to control both smuggling activities and routes effectively. In the past two decades, the leadership controlling these ethnic-based criminal networks has converted their resulting wealth into military power and political influence (Lacher, 2012). As a result of this conversion, criminal activities in the region have increasingly become militarized (Lacher, 2012). Furthermore, the technological advancements resulting in the introduction of modern infrastructure for trafficking (including means for navigation, transportation, and communication) have had disruptive effects on ethnic groups, providing further causes for conflict (Harmon, 2014).

Not surprisingly, the region traditionally hosts a wide range of different non-state armed groups. The combination of bad governance, social segregation and marginalization, poverty, crime and corruption provide fertile ground for any armed group that promises the region's disempowered population some form of change (Harmon, 2014). Many of these armed groups are rooted in and organized along local ethnic networks. Because of the many armed groups and weak state structures, the region has hosted a range of security initiatives from outside intended to stabilize the local security situation.

2.2.1 AQ's Goal and Doctrine

JNIM is firmly linked to both AQ and AQIM. Therefore, reviewing these organizations' declared goals and strategies will reveal relevant contextual information. AQ seeks to establish a Caliphate according to the methodology of the Prophet (Al-Zawahiri, 2017). It views its struggle to do so as a protracted war with the United States and Israel as its main antagonists, allied with other (primarily) Western "crusader" nations, and supported by the apostate regimes oppressing Muslim communities (Al-Zawahiri, 2017). This war "cannot be fought on a regional level without taking into account the global hostility" against the Muslim community, implying a view that this protracted war is being waged on a global level, without the restriction of national borders (Al Qaeda, 2006, p.21). Its strategy is based on the ideas of Mao, phasing its protracted war in a "Strategic Defence", "Strategic Balance" and "Decisive Phase" (Cigar, 2009).

In phase one, the organization's aim is to prevent its destruction, "smash the prestige of the regime" and signal to the population that the government is unable to prevent military attacks (Cigar, 2009, p.95). These attacks are intended to be sparse, focus on "clean targets" (i.e., Jewish and Christian targets), be widespread and dispersed over the country to overextend and disperse the defender's efforts. Some of these attacks needed to be spectacular in order to create the desired media effects. Media efforts by the group focus on establishing themselves as an alternate source of information, make clear what the struggle is about, prepare the population for mobilization and exploit the effects of spectacular attacks. Bases during this phase are lightly equipped, mobile and certainly not stationary (Cigar, 2009).

In phase two, the aim of the group is to set up larger, conventional forces to liberate areas where the regime is vulnerable or weak and replace it. In these areas, the insurgents set up administrative installations, base camps, hospitals, sharia courts and broadcasting stations. These safe areas thus also serve as jump-off points for additional military and political actions. Media efforts by the group are intended to focus on agitation of the local population by revealing the collaborationist regime's inability to defeat the insurgency. Furthermore, the group wishes to address the populations of foreign nations and convey the message that their governments have embroiled them in wars that they have no business being involved in in order to isolate the apostate regime from its foreign supporters (Cigar, 2009).

In phase three, the group's goal is to ensure that the regime is in its final phase of existence, withdrawing from the countryside into its principal cities. The group wants to see the regime suffering from political and military collapse, with internal struggles between different political factions or between politicians and the military. The latter situation potentially leads to coups. However, a total collapse is not expected as long as the regime is still receiving foreign support. The insurgents aim to make use of large scale defections to increase their own numbers, and plan to start attacking smaller cities. Once such a city falls, this is expected to start a domino-effect, eventually leading to the fall of the principal cities. Meanwhile, the insurgent vanguard intensifies cooperation with mujahedin outside of their area of operations and pursue *jihad* elsewhere once the regime collapses (Cigar, 2009).

2.2.2 AQ and AQIM's Strategy in the Sahara-Sahel Region

AQ views Mali and other countries in the Sahara-Sahel region as vulnerable and susceptible to its operations due to its near-failed state status, ongoing ethnic strife, and a large percentage of the population that is disaffected by the government (Forbes, 2018). Both AQ and AQIM have clear goals for the Sahara-Sahel region. In 2013, then AQ leader al-Zawahiri issued a global directive in which he explained that establishing unity of effort, cultivating local support, and mobilizing populations were necessary to build up a global jihadi movement as a prelude to the eventual creation of a caliphate (Al-Zawahiri, 2017). Concerning the Sahara and Sahel, AQ's strategic goal was to create an Islamic state by combining local grievances with its broader, global strategic ambition of creating a caliphate (Forbes, 2018). In 2013, after having hijacked the 2012 secular Tuareg rebellion, then AQIM's leader Droukdel declared that 'gaining a region under our control, and a people fighting for us, and a refuge for our members that allows us to move forward with our program at this stage is no small thing and nothing to be underestimated' (Forbes, 2018). In line with al-Zawahiri's guidance, he suggested to his subordinates that the 2012 Tuareg rebellion in northern Mali had presented a historic opportunity to establish a long-term presence in the Sahel (Forbes, 2018). In order to do so, Droukdel outlined five goals: uniting the Azawad people, regulating the relationship with the regional (Tuareg) armed group Ansar e-Dine, curbing the radical activities of militants in the rebellion, imposing sharia law, and developing support for AQ's activities outside the region.

2.3 JNIM's Origins, Goals, and Strategy

JNIM's existence was first announced on 02 March 2017 in a video message when its creation was reported (Zimmerer, 2022; Pollicini, 2021). JNIM is a merger of four of the most powerful armed groups in the Sahel-Sahara region that were already active in the Sahara-Sahel region, namely the Sahara Branch of Al-Qaeda in the

Islamic Maghreb (AQIM), Ansar Dine, Al-Mourabitoun, and the Katiba Macina.[1] Each of these Salafist extremist groups can be associated with one of the region's four most influential ethnic groups. In the same video message, it was declared that the Tuareg Iyad ag-Ghali was to lead JNIM and that the organization had pledged allegiance to both the AQIM and AQ leadership, as well as the Taliban.

However, JNIM operates autonomously regarding financing, tactics, and local politics (Pollicini, 2021). Some argue that the group's association with AQ mainly strengthens JNIM's authority and image rather than implying a cohesive entity (Zimmerer, 2022). Indeed, JNIM comprises several distinct Salafi-jihadist groups, each with its own historical background and unique identity. As a result, JNIM's organizational structure has been likened by some to that of a "business association," conveying the impression of a unified entity while concealing the intricate local dynamics that drive the constituent groups within JNIM. However, from an insurgency-focused perspective, JNIM has a decentralized organizational structure with a low to moderate level of central control by AQ (Jones, 2017). This is a logical organizational structure considering the circumstances in which JNIM operates, its strategy, and its function in AQ's overall strategy.

2.4 JNIMs Embedding Within AQ

Due to its ties to the different ethnicities and the criminal affiliations of its constituents, the creation of JNIM has provided AQIM and AQ with a significant geographical and social reach within the Sahara-Sahel region. This has allowed AQ to expand its support networks, including social support for recruitment and the use of criminal networks (Provoost, 2019). In the same video, it was declared that JNIM's ultimate goal is for the whole Sahara-Sahel region to be ruled under sharia law by removing oppression and expelling non-Muslim occupiers. According to Iyad Ag Ghaly, this required the unification of Muslims against crusaders, referring to the then-still-present French and MINUSMA missions (Zimmerer, 2022). JNIM justifies its use of violence based on the struggle for the freedom of oppressed Muslims (Pollicini, 2021).

While JNIM has pledged loyalty to AQ, it is assessed to operate autonomously in terms of financing, tactics, and local politics (Pollicini, 2021). JNIM is, however, firmly linked to both AQ and AQIM. Therefore, reviewing these organizations' declared goals and strategies will reveal relevant contextual information. AQ views Mali as vulnerable and susceptible to its operations due to its near-failed state status,

---

[1] Although JNIM was officially established later in March 2017, our dataset starts in January 2011. This is because the dataset contains information about the individual sub-organizations that later merged into JNIM. These groups were already active and are likely to have coordinated their operations well before the official establishment of JNIM. By including data from this earlier period, we provide a more comprehensive picture of the networks and operational patterns that led to the formal establishment of the group.

ongoing ethnic strife, and a large percentage of the population that is disaffected with the government (Forbes, 2018). Both AQ and AQIM have clear goals for the Sahara-Sahel region. In 2013, then AQ leader al-Zawahiri issued a global directive in which he explained that establishing unity of effort, cultivating local support, and mobilizing populations were necessary to build up a global jihadi movement as a prelude to the eventual creation of a caliphate (Forbes, 2018). Concerning the Sahara and Sahel, AQ's strategic goal was to create an Islamic state by combining local grievances with its broader, global strategic ambition of creating a caliphate (Forbes, 2018). In 2013, after having hijacked the 2012 secular Tuareg rebellion, then AQIM's leader Droukdel declared that 'gaining a region under our control, and a people fighting for us, and a refuge for our members that allows us to move forward with our program at this stage is no small thing and nothing to be underestimated' (Forbes, 2018). In line with al-Zawahiri's guidance, he suggested to his subordinates that the 2012 Tuareg rebellion in northern Mali had presented a historic opportunity to establish a long-term presence in the Sahel (Forbes, 2018). In order to do so, Droukdel outlined five goals: uniting the Azawad people, regulating the relationship with the regional (Tuareg) armed group Ansar e-Dine, curbing the radical activities of militants in the rebellion, imposing sharia law, and developing support for AQ's activities outside the region.

By incorporating the Tuareg and Fulani struggle for autonomy and independence into its own strategy of establishing a caliphate, AQIM is clearly operating within al-Zawahiri's guidance of 2013. Moreover, it is claimed that the creation of JNIM was a carefully prepared move by AQIM, again in line with al-Zawahiri's guidance, to develop unity of effort and thus 'create an organized, united, ideological and aware jihadi force' (Forbes, 2018). JNIM's establishment addressed AQIM's goal of uniting the Azawad people[2], as well as regulating relations with Ansar Dine and curbing the latter's radical activities. JNIM itself claimed that the merger was inspired by the 'togetherness and unification' demonstrated by factions in Syria in early 2017. Whatever the underlying motivation was, JNIM unites armed groups of different West African ethnic backgrounds (Arab, Tuareg, and Fulani) under the leadership of local Tuareg Iyad Ag Ghali while pledging allegiance to both AQIM and AQ. It is likely that this move ended the factionalism that long characterized jihadi armed organizations in the Sahara-Sahel region and that it brought four powerful armed groups under the control of AQIM and AQ. This move was facilitated by the alleged deaths of the wayward AQIM sub-commanders Belmokhtar and Abou Zeid, whom Droukdel reportedly viewed as "impetuous fanatics who were likely to alienate the very people they were trying to win over" (Forbes, 2018). This enabled AQIM to create efficiencies, promote cohesion amongst the different armed groups, and develop a more operational outlook (Forbes, 2018). As discussed, JNIM's creation also gave AQIM and AQ unprecedented geographical and social reach within the Sahara-Sahel region while expanding its support networks.

---

[2] In this context, the term 'Azawad people' refers to the diverse groups living in the Azawad region of northern Mali, united by a shared geography but divided by ethnicity and political aspirations.

When reviewing AQ's other guidelines and AQIM's subsequent goals for the Sahara and Sahel, it has been argued that JNIM achieves AQ's goal of cultivating support for the eventual implementation of Sharia law by putting much effort into religious education, as well as outreach from its religious leaders, and the provision of legal services to locals (Forbes, 2018). The goal of implementing Sharia law is now being pursued through gradual evolution instead of the radical implementation used in 2012. This is in line with AQIM's goal of curbing the radical activities of militants. It is claimed that this approach has also been enabled by the (alleged) deaths of Belmokhtar and Abou Zeid and their replacement by more religiously oriented leaders such as al-Hammam (Forbes, 2018). As a result, part of the population in JNIM's areas of operations is likely becoming accustomed to AQ's version of Sharia law and its eventual implementation (Forbes, 2018).

In its fifth regional goal, AQIM addressed Al-Zawahiri's third guideline for mobilizing populations to support AQ's efforts. Guidelines were issued that members need to patiently set conditions for locals to embrace AQ's external activities, as well as to pretend to be a domestic movement and not immediately show that AQ had global, expansionist ambitions for the whole region (Forbes, 2018). In this light, appointing Ag Ghali as the leader of JNIM was a political move to mobilize further popular support.

Apparently, AQIM's strategy for the Sahel and Sahara has been successful. In 2017, AQ issued a statement in which it was claimed that AQIM's efforts in the Maghreb and Sahel would serve as an 'epic chapter in the war annals of Muslim history' and that the unity of effort in the Sahara-Sahel region was an 'example worthy of emulation' (Forbes, 2018).

2.5   JNIMs Area of Operations

With Ansar Dine and the Katiba Macina as its largest constituents, JNIM was initially primarily active in the central and northern parts of Mali. Mali's remote, sparsely populated northeast is likely viewed as a safe area by the organization and thus of strategic importance, as it is most likely where JNIM has established its training camps, prepares for operations, and coordinates administrative tasks (Zimmerer, 2022). The areas of Timbuktu, Kidal, and Mopti are primarily relevant to the group because of the possibilities for kidnappings and drug trade that they provide (Pollicini, 2021). However, JNIM has also expanded its operations further south to Niger and Burkina Faso (Zimmerer, 2022). The systematic approach with which operations have been expanded leads analysts to suggest a strategic and tactical maturity on the side of JNIM and that it can rely on a robust local support network (ACLED, 2019). In another move signaling potential expansion, in November 2018, Kouffa, the leader of the Katiba Macina, greeted all Fulani mujahideen across the region and called on all Fulani across West Africa and Cameroon to wage Jihad. This could support the hypothesis that the increasing insecurity in Burkina Faso is part of a broader regional insurgency in the West African Sahara-Sahel region. The

2.5 JNIMs Area of Operations

steady pace at which AQ implements the strategic goals it established in 2013 is claimed to be a sign of disciplined strategic patience that might potentially enable the organization to establish a durable long-term presence in the region (Forbes, 2018). As a result of the activities of its subordinate groups, JNIM is now active throughout almost the whole of the West African Sahara-Sahel region and seems intent on expanding its area of operations even further. By doing so, JNIM managed to overstretch French and other counterterrorism forces. However, despite the expansion of its area of operations, the northern part of Mali will likely remain JNIM's organizational base due to JNIM's ethnic ties, favorable terrain, and the absence of state control. Other arguments are the lucrative border between Mali and Algeria, its location outside the footprint of counterterrorism forces, and the large percentage of the population disaffected with the government (Forbes, 2018).

JNIM interacts with several other non-state armed groups in its area of operations. One is Ansaroul Islam (AI), formerly led by Ibrahim Dicko. Dicko was a former combatant of *Mouvement pour l'unicité et le jihad en Afrique de l'Ouest* (MUJAO, which merged with Belmokhtar's al-Muthamun to form al-Mourabitoun) and allegedly was influenced by Kouffa to turn AI from an ideological network into an armed group (Raineri & Strazzari, 2017). After Ibrahim Dicko's death of natural causes, he was succeeded by his younger brother Jafar Dicko. AI is predominantly active in Burkina Faso. Another armed group that JNIM interacts with is the Islamic State in the Greater Sahara (ISGS), led by Abou Walid al-Sahrawi. Al-Sahrawi is also a former member of MUJAO. In 2015, prior to al-Murabitoun's unification (under the leadership of Belmokhtar) with AQIM, al-Sahrawi founded ISGS and pledged his allegiance to the Islamic State (IS). ISGS is predominantly active in the Mali-Niger border area. Both al-Sahrawi and Dicko have condemned the alignment of jihadist armed groups under the control of AQ (Raineri & Strazzari, 2017). Nevertheless, AI receives different forms of support from JNIM. As for ISGS, in 2018, a spokesperson claimed that JNIM shared their goal of defending Islam and hinted that collaboration could take place. It has been claimed that initially, JNIM and ISGS have conducted at least three attacks together (United Nations Security Council, 2019). More recently, however, JNIM and ISGS became rivals in establishing Salafi control over the Sahara-Sahel region. As a result, the former group's expansion into southeastern Mali and across the Burkina Faso border increased violent confrontations with ISGS (Pollicini, 2021).

The interaction with both states as well as other NSAGs, both cooperative as well as competitive, indicates that JNIM is a powerful and influential armed group in the western Sahara-Sahel region. It is argued that JNIM is a vehicle through which AQIM has sought to reassert its presence in the area it controlled prior to the French intervention in Mali in 2013 (Cristiani, 2017). Taking a more nuanced approach, it is also convincingly argued that JNIM is less of a vehicle of AQ and more an organization of local political entrepreneurs challenging local established power structures using the support of AQ (Thurston, 2020).

2.6 JNIM as an Insurgent Armed Group

Despite its designation as a Foreign Terrorist Organization (FTO) by the U.S. government, JNIM's declared goals, size, organization, composition, targets, and operations all seem to indicate that labeling it as an insurgent armed group is more appropriate than as a terrorist one (Provoost, 2019). JNIM has broad political goals, aiming to establish a territory ruled by sharia law. Much effort has been put into cultivating support for its eventual implementation through religious education, as well as through outreach from its religious leaders and the provision of legal services to locals. Furthermore, JNIM is considerably larger than a typical terrorist organization. Its unit structure is not based on cells but on fielded maneuver units. According to the available data and in line with its messaging, JNIM mainly targets security forces, though the organization is not shy of deliberately targeting civilians. However, based on the available literature and empirical evidence, its activities can be categorized as being part of the operations of a non-state armed group, asymmetrically engaging its adversary through an irregular form of warfare. Within its area of operations, JNIM uses a political-religious dominated organizational structure capable of generating income, administering territory and directing irregular armed forces, all with the aim of subverting the legitimacy of the constituted governments, ultimately replacing them to gain control over both people and resources. The next three paragraphs will elaborate on these three types of activities.

2.6.1 Use of Violence

To achieve its military goals, JNIM uses a combination of terrorism and guerrilla tactics. The organization currently promotes guerrilla action against security forces rather than attacks against civilians (Pollicini, 2021; Zimmerer, 2022). This has resulted in numerous hit-and-run attacks against the Malian Defence and Security Forces (MDSF), the Joint Force of the Group of Five for the Sahel (G5 Sahel), the United Nations Multidimensional Integrated Stabilization Mission (MINUSMA), and French forces. This focus on security forces undermines the legitimacy of the Malian government and those who support it, but it also sabotages law enforcement (Pollicini, 2021). The effect of the latter provides an environment in which criminal enterprises can thrive. Furthermore, targeting security forces aligns with AQ's broader global effort to brand itself as more moderate than competing IS-affiliates who engage more often in atrocities of different types against civilians. However, despite its claims of focusing its targeting against security forces, JNIM's violent actions have increasingly caused civilian casualties, and this increase has coincided with the expansion of JNIM's activities to Burkina Faso. This increase might be a means of responding to competition with other non-state armed groups, as well as a result of an increase in attacks against security forces, resulting in an increase in collateral civilian casualties.

Prior to focusing on guerrilla attacks against security forces, JNIM has also conducted several high-profile terrorist attacks on symbolic targets. These attacks generated much media attention, which was exploited by issuing public statements linking the attacks to the overarching strategy of the group. In particular, al-Murabitoun launched several high-profile attacks that generated publicity (Filiu, 2017; Hansen, 2019; Thurston, 2020). All these attacks can be appropriately labeled terrorist attacks, as they are politically motivated attacks that were intended to generate a psychological impact beyond the immediate victims. According to the perpetrators' media releases, the attacks had classic short-term terrorist goals like gaining publicity to mark specific events or to exact vengeance. However, the number of high-profile terrorist attacks with indiscriminate violence against civilians seems to have dropped after JNIM's creation, while in contrast, the number of attacks against security forces has increased dramatically (ACSS, 2019).

Insurgent armed groups quite commonly and deliberately use terrorism as a tactic, stopping its use once it is no longer needed or has lost its utility to achieve specific goals. In the case of JNIM, it seems that terrorism was used as an asymmetric tactic of low-level violence while the insurgency was in what Mao called the phase of *strategic defense* (Tse-Tsung 1967, p.34). The purpose of its use then was to spread disorder, gain publicity, and occasionally install vengeance, thus giving the different insurgent armed groups the time and space to concentrate on survival and political organization after the French intervention in 2013. The visible intensification of guerrilla warfare since then, leading to the creation of JNIM and the subsequent escalation and expansion of violence on a regional scale, can be seen as a sign that the insurgency has moved forward into what Mao called the phase of *strategic stalemate*.

2.6.2 Funding

Engaging in organized crime is often used to fund an insurgency (Williams, 2011). In the Sahara-Sahel region, smuggling, kidnapping for ransom, and transnational organized crime, in general, generate financial resources for armed groups (United Nations Security Council, 2019). JNIM, with its subordinate units, is no exception to this rule (Beevor, 2022; Nsaibia et al., 2023). This is hardly surprising because drug trafficking and kidnapping for ransom have become essential resources for financing the power struggles in the Sahara-Sahel region (Crisis Group Africa, 2018).

JNIM's finances initially relied primarily on kidnappings for ransom (Pollicini, 2021). Targets of kidnappings included aid workers, tourists, political actors, or people with Western nationality. However, as drug trafficking has become the essential resource for the struggles that define political power relations in the region, good relations with drug trafficking networks are essential for any politically driven non-state armed group (Crisis Group Africa, 2018). JNIM constituent al-Mourabitoun is claimed to focus on kidnapping for ransom and the trafficking of narcotics, weapons, and gasoline for financial gains (United Nations Security

Council, 2019). Al-Mourabitoun is also mentioned as having strong ties to drug traffickers transporting cocaine through the Sahel (Filiu, 2017). In particular, one of its constituent organizations, MUJAO, allegedly had a prominent role in the drug trade earlier in the decade (Harmon, 2014). This can be explained through MUJAO's roots in the Arab community, which historically controls the criminal networks (and especially the drug trade) in the Sahel (Harmon, 2014).

It has been argued that securing routes between the different drug trafficking hubs also generates significant economic activity (Crisis Group Africa, 2018). As a result, al-Qaeda affiliates, in general, have steadily become more involved in drug trafficking (Zimmerer, 2022). However, there is little (publicly available) empirical evidence for the direct involvement of AQIM and its subordinate units in the smuggling of drugs. It is more likely that the organization imposes transit taxation and provides security escorts (Boeke, 2016). For example, people smugglers who use these same routes are claimed to also be subjected to taxation by armed groups (United Nations Security Council, 2019).

Thus, JNIM in general became a more strategic criminal actor in the region by developing relationships with illicit economies across the Sahel (Beevor, 2022). Within these relationships, JNIM facilitates illicit activities by opening up illicit economies to local communities to gain support and is a direct actor, as illicit activities also form a source of funding (Beevor, 2022).

Because control over trafficking has become essential for those who wish to gain or consolidate power, it has been argued that some clashes between armed groups in the Sahara-Sahel can be traced back to disputes over trafficking and control over smuggling routes (Crisis Group Africa, 2018). Interestingly, control over smuggling routes apparently only leads to serious conflict between armed groups under certain circumstances, such as the personal involvement of a senior leader of an armed group or issues related to the overall strategic level control of smuggling routes (Crisis Group Africa, 2018). Other than under these circumstances, clashes over trafficking seem to have little impact on the general political balance of power. This indicates that there is a certain degree of deconfliction and separation of interests. Regardless of political affiliation, business interest comes first.

It is claimed that the local roots of JNIM and its Arab and Tuareg leaders are reflected in how the group positions itself concerning drug trafficking (Crisis Group Africa, 2018). Not all leaders are committed on a strictly religious basis, and they sometimes retain their past relations and interest in drug trafficking. However, as with AQIM, there is no evidence that JNIM's most influential leaders are directly involved in drug trafficking for the sake of personal gain. Some sources do claim the direct involvement of members of secondary circles around JNIM's leadership and close relatives of the group's leaders (Crisis Group Africa, 2018).

JNIM allegedly also interacts in a more operationally oriented way with criminal networks. It has been argued that JNIM cooperates with local armed bandits in Burkina Faso (Beevor, 2022). Existing criminal networks are co-opted by the insurgents with heavy weaponry and hard currency to facilitate entrance into areas with a limited support base and presence, thus expanding JNIM's area of operations. On the other hand, criminal armed groups provide manpower and logistical support to the insurgents, enabling their presence in relatively unexplored territory. As an additional benefit, siding with JNIM also provides criminal armed groups with a means of moral justification for criminal activities (Crisis Group Africa, 2018).

Despite JNIM's considerable involvement in criminal activities, there are no indications that profit trumps politics at the strategic and operational levels (Provoost, 2019). JNIM is likely involved in criminal activities as a method to fund its insurgency. However, insurgent groups are not homogenous, so one may not exclude financial gain as a primary motivation at the lower echelons. Furthermore, economic greed may go hand in hand with various other socio-economic and political grievances. However, JNIM's closed and secretive stance makes it impossible to validate such an assumption. It will be interesting to monitor whether JNIM's objectives are not corrupted by the profitability of its involvement in criminal activities over time. For now, it seems that this involvement is part of its strategy and in support of its political goals.

2.6.3 *Political Activities*

Like other AQ affiliates, JNIM has created a parallel governance system, which serves as an alternative model to state government, in which large parts of the population feel neglected (Pollicini, 2021; Nsaibia et al., 2023). By presenting itself as the defender of the people, JNIM has garnered popularity among locals. Some of the population now serve as informants that provide the group with reconnaissance information prior to attacks. In addition, by using its local social ties, JNIM's members get shelter and provisions (Pollicini, 2021; Zimmerer, 2022).

JNIM's media outlet, al-Zalaqa, is important to the group for gaining and maintaining this popular support. The platform's messages are primarily ideological, concerning retribution and "us versus them" or "good versus evil" narratives (Zimmerer, 2022). Propaganda messages include calls for violence against French and Malian forces. On Telegram, JNIM encourages civilians to resist government authority and boycott elections (Zimmerer, 2022). In addition, theologians have been publishing treatises commissioned by the group, which may help to maintain popular support by, for example, propagating the avoidance of harsh sharia punishments (Zenn, 2022).

2.7 Conclusion

In conclusion, JNIM conducts a range of administrative, funding, and violent activities. These activities can all be linked directly or indirectly to JNIM's strategy and declared political goals. Its use of terrorism seems to have been a deliberate tactic drawn from Mao's phase of *strategic defense*. Currently, JNIM's insurgent activities are those of a classic guerrilla warfare campaign, seemingly indicating a move to

Mao's phase of *strategic stalemate*. JNIM's involvement in organized crime seems to reflect the classic method of funding an insurgency. The type of criminal activities is dictated by the opportunities that the Sahara-Sahel region provides, *in casu* drug trafficking, smuggling, and kidnapping for ransom. At this stage, none of the terrorist, insurgent, or criminal activities seem to undermine JNIM's strategy or declared political goals.

References

ACLED (2019). JNIM: A Rising Threat to Stability in the Sahel. In: ACLED. https://acleddata.com/2019/02/01/jnim-arising-threat-to-stability-in-the-sahel/
Al-Zawahiri, A., (2017). Messages from the Front Lines, As Sahab Media Foundation.
Al Qaeda (2006). Yearbook: The 2006 Messages from Al Qaeda Leadership. Lulu. com, 2007.
Armed Conflict Location & Event Data Project (ACLED) (2023). *Actor profile: Jama'at Nusrat al-Islam wal-Muslimin (JNIM)*. Retrieved September 2024.
ACSS. (2019). The complex and growing threat of militant islamist groups in the Sahel.
Baldauf, S. (2001). The Cave Man and Al-Qaeda. The Christian Science Monitor, 31.
Bailes, A.J.K. & Nord, D., (2010). Non-State Actors in Conflict: A Challenge for Policy and for Law in: Mulaj, K. (ed.), (2010), Violent Non-State Actors in World Politics, New York: Columbia University Press.
Beevor, E. (2022). *JNIM in Burkina Faso: A Strategic Criminal Actor*. Global Initiative Against Transnational Organized Crime.
Boeke, S. (2016). Al Qaeda in the Islamic Maghreb: Terrorism, insurgency, or organized crime? *Small Wars & Insurgencies, 27*, 5.
Bunker (2005). Networks, Terrorism and Global Insurgency, Routledge, London.
Cigar, N. (Ed.). (2009). Al-Qa'ida's Doctrine for Insurgency: Abd Al-Aziz Al-Muqrin's A Practical Course for Guerrilla War. Potomac Books, Inc.
Crsis Group Africa. (2018). *Report N°267: Drug trafficking, violence and politics in northern Mali*. International Crisis Group.
Cristiani, D., (2017). Ten Years of al-Qaeda in the Islamic Maghreb: Evolution and Prospects, in: Terrorism Monitor, Jamestown Foundation
Forbes, J. (2018). Revisiting the Mali Al-Qa'Ida playbook: How the group is advancing on its goals in the Sahel. *CTC Sentinel, 11*, 9.
Filiu, J-P., (2017). Al-Qaida in the Islamic Maghreb and the Dilemmas of Jihadi Loyalty, in: Perspectives on Terrorism, Vol.11 (6)
Guidère, M. (2014). The Timbuktu letters: News insights about AQIM. *Res Militaris, 4*, 1.
Hansen, S. J. (2019). *Horn, Sahel and rift fault-lines of the African Jihad*. Hurst & Company.
Harmon, S. A. (2014). *Terror and insurgency in the Sahara-Sahel region: Corruption, contraband, jihad and the Mali war of 2012–2013*. Ashgate Publishing.
Jones, S. G. (2017). *Waging insurgent warfare*. Oxford University Press.
Lacher, W., (2012). Organized Crime and Conflict in the Sahel-Sahara Region, Washington: Carnegie Endowment for International Peace Publications Department.
Ministers van Buitenlandse Zaken, van Defensie, voor Buitenlandse Handel en Ontwikkelingssamenwerking en van Veiligheid en Justitie nr. 213, (2018). Het Besluit van het Kabinet een Nederlandse Bijdrage te leveren aan de UN Multidimensional Integrated Stabilisation Mission in Mali (MINUSMA), Tweede Kamer, vergaderjaar 2017–2018, 29251 (nr.368).
Mulaj, K. (2010). *Violent non-state actors in world politics*. Hurst.

References

Nsaibia, H., Beevor, E., & Berger, F. (2023). *Non-state armed groups and illicit economies in West Africa: Jama'at Nusrat al-Islam wal-Muslimin (JNIM)*. Global Initiative against Transnational Organised Crime (GI-TOC).

Provoost (2019). Symbiosis in the Sahara and Sahel : why and how does Jama'a Nusrat ul-Islam wa al-Muslimin' (JNIM) combine insurgent, terrorist and criminal activities and what does this reveal about the movement's true nature?, Netherlands Defence Academy, Breda.

Pollicini, L. (2021). A case of violent corruption: JNIM'S insurgency in Mali (2017-2019). *Small Wars & Insurgencies, 32*(7), 1092–1116.

Raineri, L., Strazzari, F., (2017). Jihadism in Mali and the Sahel: Evolving dynamics and patterns, in: Brief 21, European Union Institute for Security Studies (EUISS).

Reitano, T., Clarke, C., & Adal, L. (2017). *Examining the nexus between organised crime and terrorism and its implications for EU programming*. CT MORSE.

Thurston, A. (2020). *Jihadists of North Africa and the Sahel: Local politics and rebel groups*. Cambridge University Press.

Tse-Tsung, M. (1967). *On protracted war. Translated by foreign languages press. Third edition (second printing)*. People's Publishing House.

United Nations Security Council. (2019). S/2019/50: Twenty-third report of the Analytical Support and Sanctions Monitoring Team submitted pursuant to resolution 2368 (2017) concerning ISIL (Da'esh), Al-Qaida and associated individuals and entities.

Williams, P. (2011). Criminals, militias, and insurgents: Organized crime in Iraq. In D. P. Branson & A. J. Mitchell (Eds.), *Crime and insurgency in Iraq and Afghanistan*. Nova Science Publishers.

Zenn, J. (2022). Rebel rivals: Reinterrogating the movement for Unity and Jihad in West Africa and the roots of Al-Qaeda-Islamic state infighting in the Sahel. *The Journal of North African Studies, 27*(6), 1277–1301.

Zimmerer, M. (2022). Terror in West Africa: A threat assessment of the new Al Qaeda affiliate in Mali. *Critical Studies on Terrorism, 12*(3), 491–511.

Chapter 3
Temporal Probabilistic Rules and Policy Computation Algorithms

The principal goal of this book is to develop a model that is capable of predicting attacks by JNIM—just as has been done previously in the case of Lashkar-e- Taiba (Subrahmanian et al., 2012), the Indian Mujahideen (Subrahmanian et al., 2013), and Boko Haram (Subrahmanian et al., 2020). Chapter 1 of this book shows that the predictive models used are highly accurate in predicting many of the types of attacks that JNIM carries out. However, as stated in Subrahmanian & Kumar (2017), a good predictive model must have three components:

First, the predictive model must be *accurate*—it must make predictions that are correct most of the time. It is important to note that accuracy is measured in machine learning through a variety of technical metrics. These include:

- Precision/Confidence: Of all predictions that an attack will happen, precision is the percentage of times the attack did in fact happen during the time frame of interest.
- Recall: Of all attacks that did occur during the time frame of interest, recall refers to the percentage of those attacks that were predicted to occur.
- F1-Score: There are many applications where the precision is very high and the recall is very low. The F1-score combines the previous two metrics by computing their harmonic mean.
- Accuracy: The technical definition of accuracy simply computes the percent-age of predictions made (attack will occur, and attack will not occur) that agree with what actually occurred or did not occur during the time frame in question. As a metric, accuracy can be wildly inappropriate. For example, if an attack only occurs in 10 of 100 months, a predictor that predicts that that attack will never occur will have an accuracy of 90%—but would have 0% recall and would never predict a single attack.
- Other metrics for measuring predictive accuracy include the popular Area Under a ROC Curve.

All of these metrics lie in the [0,1] interval with a higher number denoting better predictive performance. To achieve high accuracy with respect to these metrics, the Northwestern Terror Early Warning System (or NTEWS) discussed in Chap. 1 uses an ensemble of six machine learning classification algorithms that are combined together using a technique called late fusion. Because such classification algorithms are well-known and widely used in the literature, we do not discuss their details here.

Second, the predictive model's predictions must be *explainable* to a person who is knowledgeable in the field in which the prediction is being made. In the case of this book, that means that the predictive models' predictions must be explainable to a counterterrorism or a military or intelligence official dealing with JNIM in the Sahel region. Such individuals are well trained in their field but are usually not experts in computer science. The Temporal Probabilistic (TP) rule paradigm used in this book and described in more detail in this chapter focuses on explainability. TP-rules were first introduced in Subrahmanian et al. (2012) and built upon prior work merging probabilities and time together within rules (Dekhtyar et al., 1999).

Third, the predictive model must provide enough insight to enable *actionability*. An expert on JNIM not only needs to know what a predictive model of JNIM predicts they might do in the coming six months. It also needs to provide suggestions on how JNIM's attacks might be mitigated. Such suggestions of strategic actions will enable policy makers to shape actions on the ground that reduce the intensity of JNIM's attacks. We will discuss a policy computation algorithm later in this chapter.

3.1 JNIM Data

Our JNIM dataset is represented as a relational table—for non-computer scientists, the tables are just spreadsheets. A *row* in the JNIM dataset corresponds to a month during the period 2011–2022 (12 years in all). Though "12 years of data" may sound like a lot to some, for computer scientists, this dataset consisting of 144 rows is in fact very small. The columns in the JNIM dataset consist of two types of variables:

- *Dependent (Attack) Variables:* These are the types of attacks we want to predict. In this book, we look at predicting JNIM attacks 1, 2, 3, 4, 5, 6 months into the future. In other words, this book provides methods to predict—say on Jan 1, 2020—the attacks that will happen in Jan 2020, sometime in the Jan 2020—Feb 2020 period, and so forth until and including the Jan 2020—June 2020 period. The same techniques described in this section can be easily adapted to make such predictions in other periods of time (say for the next year).
- *Independent (Environmental) Variables:* These variables capture various aspects of the context in which JNIM operates. They include information that falls into three broad categories:

- *Actions that JNIM takes*. Some actions that JNIM takes such as addressing the Malian government directly in public communications, attacking public sites, and abduction of people are examples of actions taken by JNIM.
- *Social, Cultural, Political, Economic, Military Variables*. These constitute the bulk of the independent variables. These variables may include, for instance, JNIM actively pushing out messages about the religious agenda, variables linked to elections in Mali, variables related to the availability of jobs, treatment of women, and more. They may also include actions taken by third parties that affect the environment in which JNIM operates—such as raids, arrests and kills of JNIM personnel by Malian security forces, international trials or condemnation of JNIM, and more.
- *Group-related Internal Variables*. Such variables include structural information about the group such as the nature of the leadership of the group, whether there is internal dissension within the group, and the nature of the group's relationships with other armed groups.

Appendix B contains a comprehensive summary of the variables used in our JNIM dataset.

3.2 Temporal Probabilistic Rules

At a high level, a temporal probabilistic rule is an expression of the form "If an environmental condition C holds during month m, then an attack A will (or will not) occur sometime during the month m to $m + \delta$ timeframe". Thus, δ is a time delay which specifies a range of months in the future, during which the attack A is predicted to occur.

An environmental condition is a condition over the environmental variables. In the case of the JNIM data, all variables are binary (yes/no). If E is an environmental variable, then $E = 0$ and $E = 1$ are *environmental atoms*. Similarly, if A is an attack variable, then $A = 0$ and $A = 1$ are *attack atoms*. A TP-rule r is an expression of the form:

$$EA1 \& \ldots \& EAn \rightarrow AA : \delta$$

where $EA1, \ldots, EAn$ are environmental atoms and AA is attack atom, and $\delta \geq 1$ is an integer. The integer δ is the time delay mentioned above—in this book, we restrict δ to be between 1 and 6 as all predictions are made for the next 6-month period.

Despite the apparently formal statement of the definition of TP-rules given above, the rule can be easy read in English as: "*if EA1, ..., EAn are all true in a given month m*, then AA will (or will not) occur sometime during the next $m + \delta$ months". The rule is read as "attack A will occur" when the attack atom AA has the form $A = 1$ for some attack A and "attack A will not occur" when the attack atom AA has the form $A = 0$.

We call *EA*1 & ... & *EAn* the *pre-condition* of the above rule and *AA* the *conclusion* of the above rule.

For example, consider a TP-rule about JNIM that says that: "if the national government where JNIM operates instituted a travel ban and the government did not raid JNIM facilities or institutions during a month m, then JNIM will release abducted people 6 months later". In this case,

- *EA*1 is the atom "government instituted a travel ban" and
- *EA*2 is the atom "JNIM facilities were not raided" and
- *AA* is *Abduction Release* = 1 and
- $\delta = 6$.

We call such rules *positive* TP-rules because they predict an event will occur, i.e. the event atom *AA* in the conclusion of such rules are of the form $A = 1$.

Sometimes, we are interested in predicting that an event will not occur, i.e. when the attack atom *AA* in the conclusion of such rules are of the form $A = 0$. We call such rules *negative rules*.

An example of a negative rule is the rule "if the nation's security forces did not execute civilians and there was no state of emergency declared during a month m, then JNIM will not carry out attacks on security installations within the next two months". In the case of this rule:

- *EA*1 is the atom "the nation's security forces did not execute civilians" and
- *EA*2 is the atom "there was no state of emergency declared" and
- *AA* is attacks on security installations = 0 and
- $\delta = 2$.

Our temporal-probabilistic rule mining framework can easily extract such rules from data of the kind described in Sect. 3.1.

Every TP-rule *EA*1 & ... & *EAn* → *AA*: δ has 5 associated statistics that we compute.

Confidence The confidence of the above rule is the conditional probability of *AA* being true sometime within the next δ months, given that the precondition is true in month m. We want TP-rules to have high confidence.

Negative Confidence The negative confidence of the above rule is the conditional probability of *AA* being true within the next δ months, given that the precondition is false in month m. We want TP-rules to have low negative confidence. Intuitively, when a TP-rule has high confidence and low negative confidence, then the precondition *EA*1 & ... & *EAn* serves as a beacon. When it is true (turned on), it predicts that *AA* will be true within the next δ months. When it is false (turned off), it predicts that *AA* will not be true within the next δ months in the future. The bigger the "spread" (difference) between confidence and negative confidence, the better the rule distinguishes between whether *AA* is true within the next δ months in the future and when *AA* is false within the next δ months in the future.

Inverse Confidence The inverse confidence of the above rule is the conditional probability of the precondition of the above rule being true in month $(m - \delta)$, given

3.2 Temporal Probabilistic Rules

that AA is true sometime during the time interval between month $(m - \delta)$ and month m. We want TP-rules to have high inverse confidence.

Lift Sometimes a rule can have high confidence and inverse confidence because AA is almost always true—in this case, the precondition is not of much use because AA is true regardless of whether the precondition is true or not. The *lift* of a rule is the ratio of the confidence of the rule, divided by the prior unconditional probability of the conclusion (AA) of the rule. We want a lift to be greater than 1—the larger the better.

Support The support of the rule is the percentage of months m such that the complex "AND" condition $EA1$ &...&EAn is true in month m and AA is true sometime between now (i.e., month m) and month $(m + \delta)$. We want support to also be high, but at least 5% if possible. Intuitively, the support of a rule makes sure that there are sufficiently many cases in which $EA1$ & ... & EAn and AA hold in the appropriate months—that the rule is not a "one of" type of situation.

All of the above metrics lie between 0 and 1 except for lift which can have values greater than 1. We want high values for all of the above metrics except for negative confidence (which should be low)—and in the case of lift, the value should be higher than 1.

TP-rules build upon the idea of a class of rules called *association rules* (Agrawal & Srikant, 1994) which were first used to help companies like Walmart perform *market basket* analysis which seeks to understand customer buying patterns in grocery stores, department stores and more. For example, association rules can be used to identify the probability that customers who buy milk and eggs may also buy bread. Association rule *mining* algorithms try to identify rules which are guaranteed to satisfy thresholds for support and confidence. The best-known algorithms for association rule mining are the Apriori algorithm (Agrawal & Srikant, 1994), but there are many more sophisticated implementations which are far more efficient (Tan et al., 2016). However, association rules cannot capture the temporal dependencies that are necessary for reasoning about terrorist groups such as Boko Haram. TP rules, first introduced in Subrahmanian et al. (2012), can be viewed as a combination of temporal probabilistic logic programs (Dekhtyar et al., 1999) and association rules.

Mining TP-rules from a body of data can be done via specialized methods such as those in Subrahmanian & Ernst (2009) or via a simple application of association rule mining methods. Suppose we assume (as is the case with the JNIM dataset) that each row in the input table T corresponds to a month m. Then to mine association rules whose conclusion is of the form $A = j: \delta$ for $j \in \{0,1\}$ and for $\delta \in \{1, ..., 6\}$, we create a new table Tj, δ as follows.

(a) Delete all dependent variables except for the dependent variable A.
(b) For the rows corresponding to month m in T s.t. $m \leq MAX - \delta$, replace the entry in column A by the value of A sometime during the interval from month m to

month ($m + \delta$). This value is set to 1 if the attack happened sometime during the interval from month m to month ($m + \delta$)—otherwise it is set to 0. Here, *MAX* is the last month in the dataset.

(c) Delete rows ($MAX - \delta + 1, \ldots, MAX$).

For the row corresponding to month m in T, replace the entry in column A by the value of A during the interval from month m to month ($m + \delta$).

We can then apply any standard association rule mining algorithm to Tj, δ in order to get TP-rules with $A = j: \delta$ in the conclusion of the rule and with a delay of δ.

Because standard association rules typically support only confidence, support and (sometimes) lift thresholds, the resulting set of TP-rules needs to go through a further filtering step to make sure that the final set of rules returned satisfy user-specified support, confidence, inverse confidence, negative confidence, and lift requirements.

3.3 Conclusion

This chapter provided a quick overview into the ideas behind the techniques used to compute JNIM rules in the Sahel region for the period coded. Temporal probabilistic rules were introduced briefly for a broad audience. Those interested in the mathematical foundations behind these techniques are advised to consult the references as provided. Finally, we want to emphasize that correlation is not the same as causation. The rules on the behavior of JNIM provided in this book are *not* causal. Rather, they are correlations that provide decision support to subject matter experts knowledgeable in the field. We reflect on the value of this in Chap. 9.

References

Agrawal, R., & Srikant, R. (1994, September). Fast algorithms for mining association rules. In: Proceedings of the 20th international conference on very large data bases, VLDB, vol 1215, pp 487–499.

Dekhtyar, A., Dekhtyar, M., & Subrahmanian, V. S. (1999). Temporal probabilistic logic programs. In *Proceedings of 1999 international conference on logic programming, New Mexico, November* (pp. 109–123).

Independent (2019). OPPI: army arrests undercover aids of Boko Haram. Retrieved June 8, 2020, from https://www.independent.ng/oppi-army-arrests-undercover-aides-of-boko-haram/

Subrahmanian, V. S., & Ernst, J. (2009). Method and system for optimal data diagnosis. U.S. Patent 7,474,987. University of Maryland, Baltimore.

Subrahmanian, V. S., & Kumar, S. (2017). Predicting human behavior: The next Frontiers. *Science, 355*(6324), 489.

Subrahmanian, V. S., Mannes, A., Sliva, A., Shakarian, J., & Dickerson, J. (2012). *Computational analysis of terrorist groups: Lashkar-e-Taiba*. Springer.

References

Subrahmanian, V.S., Pulice, C., Brown, J.F. and Bonen-Clark, J. (2020). A machine learning based model of Boko Haram (pp. 1-116). Cham: Springer.

Subrahmanian, V. S., Mannes, A., Roul, A., & Raghavan, R. K. (2013). *Indian Mujahideen: Computational analysis and public policy*. Springer.

Tan, P. N., Steinbach, M., & Kumar, V. (2016). *Introduction to data mining*. Pearson Education India.

Chapter 4
Abduction and Release of Abductees

Abduction is defined as the 'taking away of a person away from a place by force' (see: See https://www.britannica.com/topic/abduction). The term 'kidnapping' is often used to describe the same process but expands the definition by describing the process as the unlawful taking and carrying away of a person by force or fraud or the unlawful seizure and detention of a person against his will. When coding abductions undertaken by JNIM, events were not coded as abductions when they did not fit this definition.

On a conceptual level, there has been debate within AQ about the utility of kidnapping and hostage-taking as part of an overall strategy (Cigar, 2009, p.26). In his writings, al-Muqrin argued that kidnapping might be useful if there is a rational, concrete objective for it. These objectives might be forcing a government to accede to demands, embarrassing a local government, obtaining information, generating income for the central organization or generating publicity for a specific cause (Cigar, 2009, p.156). Furthermore, Al-Muqrin differentiated between covert and overt hostage-taking, with the former being the less risky method of the two. Carefully selecting the target for an abduction was deemed important. This not only concerned intelligence on the target, but also its significance in relation to the goal of the abduction. For example, human targets were prioritized in significance in the order of Jews, Christians and apostates. In practice, the central organization of AQ has used abductions since the early 2000's both as a way to finance group activities and expansion, as well as generating publicity for its cause.

When discussing abduction in the Sahara-Sahel region, one should be aware that in some cases, criminal organizations engage in kidnappings and subsequently hand over hostages to Islamic insurgents in exchange for money. Although the initial abduction was conducted by criminals, insurgents subsequently claimed to be the kidnappers, using their hostages for ideological purposes, propaganda, financial gain or as bargaining chips in their broader strategic objectives. This opportunistic cooperation between criminal elements and insurgents underscores the networked

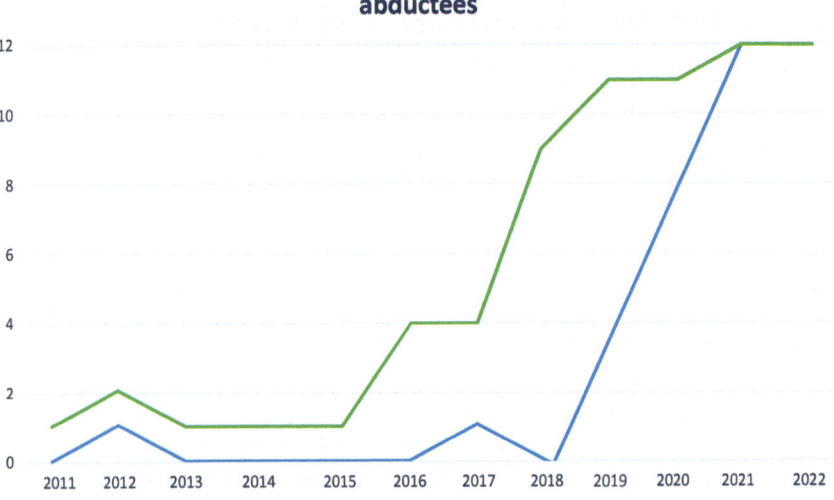

Fig. 4.1 Number of months per year when abduction occurred, or abductees were released

structure of the various NSAGs in the region, but also blurs the lines between the phenomena of organized crime and insurgency. This phenomenon may present challenges to analysts and policymakers in their endeavors to ascertain the nature of a Non-State Armed Group (NSAG) and may constrain their capacity to comprehend and assess the organization's activities. Political framing may then lead to mischaracterizing an insurgent organization as a terrorist group which utilizes religious rhetoric as a cover for engaging in profit-driven criminal activities.

As an AQ affiliate, the finances of JNIM (and the predecessors that became part of it) initially primarily relied on kidnappings for ransom. Targets of kidnappings included aid workers, tourists, political actors, or people with Western nationality. While the focus initially was on kidnapping foreigners with a Western background, the organization started targeting almost only local citizens in the past years (Berger, 2023).

During the period under our investigation, JNIM carried out abductions in 79 out of 144 months. Notably, between 2018 and 2022, there was a significant rise in abductions, with JNIM reportedly conducting abductions in 54 out of 144 months. Figure 4.1 visually represents the frequency of reported abductions during the studied timeframe and illustrates the upward trend in abduction incidents. Additionally, this figure depicts the months when JNIM released abductees, often after the JNIM collected ransom. Figure 4.2 likewise shows the number of abducted people per month as well as the number of released abductees.

As stated before, JNIM initially engaged in kidnappings primarily for financial gain, intertwined with ideological motives. Kidnapping Westerners is deemed especially fitting when the aim is to embarrass the local government, generate income and generate publicity for the Salafist-Islamic cause in the Sahara-Sahel region. Unsurprisingly, this resulted in a relatively high number of abductions of Western

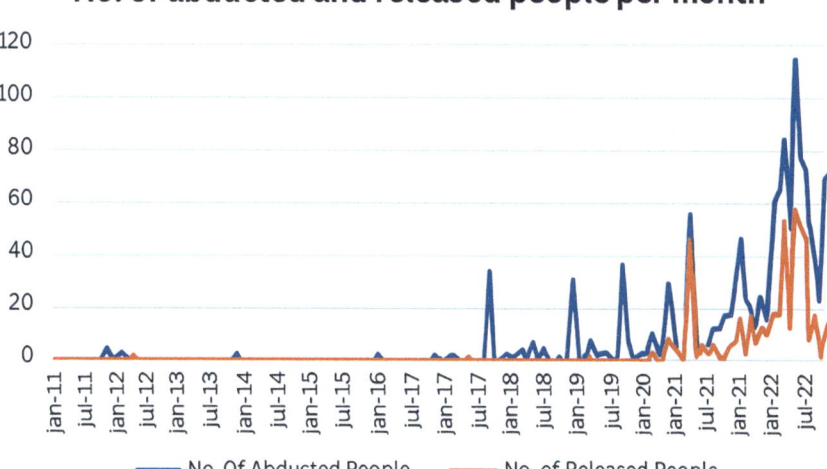

Fig. 4.2 Number of abducted and released people per month

individuals during the early years of our study period. For instance, JNIM reported that they received $19.4 million in July 2012 in exchange for releasing two kidnapped Spaniards and one Italian (Nünlist, 2013).

Following the merger of several sub-groups into JNIM in 2017, abductions apparently have also been utilized as a means to gain strategic control over the territories within their influence (Berger, 2023). Abductions of local citizens serve dual purposes, including gathering information on and intimidating the local population, all while contributing to the financial needs of JNIM. However, in 2022, there appears to have been a change in their approach, likely due to the increasing opposition faced by JNIM. Subsequently, JNIM apparently reverted to abducting individuals from Western backgrounds, using this again as a cost-effective method of generating income (Berger, 2023). The subsequent instances of abduction further illustrate the evolution in JNIMs' strategies and tactics concerning abductions.

Examples of some prominent abductions are given below.

- **23 November 2011**: A gang of armed men kidnapped two French nationals from their hotel in northeastern Mali overnight. The assailants were unknown initially, but several days later turned out to be AQIM. They were released in December 2024 in exchange for two AQIM prisoners. (AFP, 2011)
- **November 2013**: Two RFI journalists from France were abducted subsequent to concluding an interview with a regional MLA leader. Their bodies were found several miles east of Kidal, the executions were later claimed by AQIM
- **17 January 2016**: Two senior Australian doctor-missionaries were abducted by AQIM in the vicinity of Baraboulé in northern Burkina Faso, where they had resided for numerous decades. Both were released weeks after.
- **4 September 2017**: 30 people in Mamba were reportedly kidnapped by JNIM as revenge for the beating up of two Islamists after a quarrel about veil wearing

- **4 March 2019**: Presumed Katiba Macina (JNIM) militants abducted four civilians including a teacher near the village of Mountou
- **19 April 2020**: JNIM fighters reportedly abducted civilians supportive of ISGS, and set up checkpoints in the area of Ntillit (Al-Naba, 2020)
- **27 July 2020**: Suspected JNIM militants abducted the chief of a Nassoumbou village
- **April 2021**: Kidnapping of Olivier Dubois in Gao, northern Mali (French journalist) (Reuters, 2021)
- **9 September 2021**: Suspected Katiba Macina (JNIM) militants abducted an NGO worker along with a vehicle in the town of Diallassagou (Bandiagara, Mopti)
- **13 July 2022**: Presumed JNIM militants abducted about 20 women, girls, and children searching for baobab leaves in Dougouri-Ouidi

Kidnappings by JNIM have a major impact on locals and foreigners visiting Mali, Niger and Burkina Faso. However, the extent of abductions and the people being targeted can reveal some information about JNIM's greater strategic goals or the current (financial) state of the organization. The focus of this chapter is to derive TP-Rules to predict when JNIM will abduct people, as well as when the group releases (some of) its prisoners. We will now briefly discuss some of the events and variables linked to the abductions, starting with the variables linked to predicting the releasement of abductees.

- *The Malinese, Nigerien or Burkinese government reportedly ordered execution(s) of declared offenders.* In months in which the Malinese, Nigerien or Burkinese government ordered executions of declared offenders, and some other conditions were met, we were able to derive TP-rules for the next 1 month period.
- *JNIM discusses whether their campaign has been achieving the desired objectives.* The discussion of JNIM's campaign is used, with other variables, to derive several TP-rules. Using these rules, we can predict release of abductees sometime during the next 2 months.
- *A travel ban was placed on Mali, Niger, or Burkina Faso.* By examining months when a travel ban was placed by foreign countries on Mali, Niger or Burkina Faso, we were able to derive TP-rules to predict release of abductees sometime during the next 6 months.

All of the previously mentioned variables are those that occur most frequently when predicting the release of abductees. The following variables occurred the most when predicting abductions.

- *A travel ban was placed on Mali, Niger, or Burkina Faso.* By examining months when a travel ban was placed by foreign countries on Mali, Niger or Burkina Faso, we were able to derive TP-rules to predict abductions during the next 3 months.
- *A country or international organization freezes the assets of (members of) JNIM.* In months where some other conditions were met and a country or international organization froze the assets of (members of) JNIM, we were able to

derive TP-rules associated with both the next 1 month and the next 3 month windows.
- *A travel ban was placed on Mali, Niger, or Burkina Faso.* By examining months when a travel ban was placed by foreign countries on Mali, Niger or Burkina Faso, we were able to derive TP-rules in conjunction with other variables to predict abductions for the next 1 month and the next 3 month window.

4.1 Release of Abductees when the National Government Ordered Executions and there Was no State of Emergency

During a period where the nation's security forces ordered execution of declared offenders and there was no state of emergency in effect, we were able to derive a TP-rule to predict release of abducted people by JNIM sometime in the next 1 month. For example, in May of 2021, the Malian national government ordered executions of declared offenders, and there was no state of emergency. One month later in June a total of 12 abductees were released by JNIM. The results of TP-Rule AR-1 are displayed below.

4.1.1 TP-Rule AR-1

Release of abducted people by JNIM occurs in months in which:
- 1 month earlier, the national government where JNIM operates, ordered the execution of declared offenders.
- 1 month earlier, there was no state of emergency.

   Support = 0,19
   Probability = 71%, *Inverse Probability* = 69%, *Negative Probability* = 11%, *Lift* = 2.6

4.2 Abductions When a Travel Ban Was Placed on the State Where JNIM Operates [AR]

In months where a travel ban was placed on the country where JNIM operates, we were able to derive a TP-rule to predict the abduction of people by JNIM sometime in the next 3 months. Between 2011 and 2022, prolonged restrictions on travel primarily stemmed from the global impact of the COVID-19 pandemic. However, the travel restrictions under discussion here pertain specifically to limitations imposed on individuals associated with JNIM by international institutions or foreign states. In this context, these travel restrictions often form part of comprehensive sanctions or are enforced by neighboring countries of Mali, Niger, or Burkina Faso.

4.2.1 TP-Rule AR-2

Abduction of people by JNIM occurs in months in which:

- Up to 3 months earlier, a travel ban was placed on the country where JNIM operates.

  Support = 0,31
  Probability = 92%, *Inverse Probability* = 63%, *Negative Probability* = 28%, *Lift* = 1.8

4.3 Release of Abductees When a Travel Ban Was Placed on the State Where JNIM Operates and the Government Did Not Raid Facilities or Institutions [AR]

Similar to AR-2, TP-Rule AR-3 looks at months when a travel ban was placed on the country where JNIM operates, often limiting cross-border movements. However, AR-3 also considers an additional variable. In months where the national government where JNIM operates institutes a travel ban and the government did not raid JNIM's facilities or locations, we were able to derive a TP-rule to predict release of abducted people by JNIM sometime in the next 6 months. For example, in May of 2020, the Malian government instituted a travel ban and in addition, the government did not raid facilities or institutions owned by JNIM. 6 months later in October of 2020, JNIM released 33 Dozo militiamen in Markala Coura who were previously abducted. The TP-rule we derived is shown below.

4.3.1 TP-Rule AR-3

Release of abducted people by JNIM occurs in months in which:

- upto 6 months earlier, a travel ban was placed on the country where JNIM operates.
- upto 6 months earlier, the national government did not raid JNIM's facilities or institutions.

  Support = 0,19
  Probability = 55%, *Inverse Probability* = 67%, *Negative Probability* = 14%, *Lift* = 1.6

4.4 Abductions When a Travel Ban Was Placed on the State Where JNIM Operates and JNIM Was Not in Internal Conflict [AR]

We were also able to derive TP-rules predicting abductions based upon the occurrence of travel bans in combination with occurrences when JNIM was not enmeshed in internal conflict. One real life example of this rule in action could be seen in the period February to December of 2020. In this period, there was a travel ban placed on members of JNIM and there was no internal conflict while a total of 116 individuals were captured or abducted by JNIM ranging from political officials to teachers and healthcare workers. In months where a travel ban was placed on the country where JNIM operates, and JNIM was not enmeshed in internal conflict over unspecified objectives, we were able to derive a TP-rule to predict the abduction of people by JNIM 1 month in the future.

4.4.1 TP-Rule AR-4

Abduction of people by JNIM occurs in months in which:

- 1 month earlier, a travel ban was placed on the country where JNIM operates.
- 1 month earlier, JNIM was not enmeshed in internal conflict over unspecified objectives.

Support = 0,30
Probability = 91%, *Inverse Probability* = 61%, *Negative Probability* = 28%, *Lift* = 1.9

4.5 Abductions When Foreign States or International Institutions Froze Assets [AR]

Since 2015, the UN security council has been imposing sanctions on individuals and entities for obstructing the 2105 Malian peace deal. This includes individuals and entities who in 2017 became known under the name of JNIM. Several other foreign states have been following in the Security Councils' footsteps, also imposing sanctions on the same entities and individuals. In months where foreign states or international institutions froze assets of JNIM, we were able to derive a TP-rule to predict the abduction of people by JNIM 3 months in the future. TP-rule AR-4, is depicted below.

4.5.1 TP-Rule AR-5

Abduction of people by JNIM occurs in months in which:

- Up to 3 months earlier, foreign states or international institutions froze asset(s) of individual member(s) of JNIM.

  Support = 0,31
  Probability = 90%, *Inverse Probability* = 63%, *Negative Probability* = 28%, *Lift* = 1.8

4.6 Abductions When Foreign States or International Institutions Froze Assets and the Government Did Not Raid Facilities or Institutions[AR]

Like AR-5, TP-Rule AR-6 examines instances when foreign states or international institutions froze assets of individuals connected to JNIM or JNIM as an entity. However, AR-3 introduces an extra variable. In months when the national government were JNIM operates freezes the assets of JNIM, without conducting any government raids on locations or institutions connected to JNIM, we formulated a TP-rule to forecast the abduction of individuals by JNIM sometime in the next 1 month. For example, in February 2019, international institutions imposed several sanctions on JNIM members, and the government refrained from raiding JNIM-owned facilities or institutions. One month later, in March 2019, JNIM militants abducted four civilians, including a teacher, in the vicinity of Mountou, Mali.

4.6.1 TP-Rule AR-6

Abduction of people by JNIM occurs in months in which:

- 1 month earlier, foreign states or international institutions froze asset(s) of individual member(s) of JNIM.
- 1 month earlier, there were no government raids on the JNIM.

  Support = 0,30
  Probability = 91%, *Inverse Probability* = 61%, *Negative Probability* = 28%, *Lift* = 1.9

4.7 Abduction When Foreign States or International Institutions Froze Assets and Security Forces Were Not Deserting [AR]

Finally, we have also formulated TP-Rules based on the occurrence of JNIM's asset freezing coupled with the absence of security forces desertions. In months when the national government in the region where JNIM operates freezes assets belonging to JNIM, and there is no reported desertion within the security forces of that nation, we successfully derived a TP-rule capable of predicting abduction of individuals by JNIM sometime in the next 3 months. This TP-Rule (AR-7) is depicted below.

4.7.1 TP-Rule AR-7

Abduction of people by JNIM occurs in months in which:

- Up to 3 months earlier, foreign states or international institutions froze asset(s) of individual member(s) of JNIM.
- 3 months earlier, security forces were not deserting in the nation in which JNIM operates.

Support = 0,30
Probability = 93%, *Inverse Probability* = 60%, *Negative Probability* = 29%, *Lift* = 1.9

4.8 Conclusions

In recent years, JNIM has engaged in numerous abductions for various reasons, often intertwining political objectives with financial motives. Understanding the motive for abductions is crucial, not only to formulate a response and to determine a negotiating strategy, but also to gain insights into the state of the insurgent organization and its strategy. The motive behind abductions can offer valuable information regarding the insurgent group. For instance, if abduction appears primarily financially motivated, it may suggest opportunistic behavior or resource needs within the group, potentially indicating weaknesses or vulnerabilities that can be exploited by counterinsurgency efforts. Conversely, if the abduction is ideologically driven, it may signify the group's commitment to its cause, its willingness to leverage violence for ideological purposes, or a need to bring attention to its cause. However, understanding the motives behind kidnappings requires the incorporation of additional data variables into event analysis, such as media releases associated with the kidnapping and background information about the hostages. This is information that is not incorporated in the outcome of the model but can greatly enhance qualitative analysis.

Our research identified several conditions associated with occurrences of abduction:

- Abductions tend to occur in months when, sometime in the preceding 3 months prior, a travel ban was imposed on the country where JNIM operates.
- Similarly, abductions are observed to follow in the month after periods when JNIM did not experience internal conflicts.
- Abductions are also noted in months when, sometime in the preceding 3 months, the assets of individual members of JNIM were frozen by foreign states or international institutions.
- In months where the assets of individual members of JNIM are frozen by foreign entities, and the national government refrains from raiding JNIM facilities or experiencing desertions in national security forces, abductions often follow either during the next 1 month or next 3 months.

It is essential to note that these conditions are not causative but offer insights into circumstances where abductions are more prevalent. For instance, an increase in abductions can be expected when security forces raid JNIM facilities or locations.

Regarding the release of abductees:

- Abductees tend to be released within 3 months of months when the national government where JNIM operates orders the execution of declared offenders and there is no state of emergency.
- When the national government refrains from raiding JNIM locations or facilities, and a travel ban is imposed on the state where JNIM operates, the release of abductees seems to occur within the next 6 months.

Just like the conditions related to abductions, these events are not causative but provide insights into the conditions influencing the release of abductees. For example, the execution of declared offenders could positively influence the release of abductees. Additionally, a common theme emerges, indicating that both a travel ban on the state in which JNIM operates and the freezing of assets may lead to an increase in abductions.

4.9 Predictive Model/Reports Results

Table 4.1 below shows the performance of our Northwestern Terror Early Warning System's (NTEWS) predictive model. We learned the predictive model using data up to the end of 2022 and then made monthly predictions during 2023. In other

Table 4.1 Abduction of people: predictive results for our 2023 predictive reports

Abduction						
Time period	1	2	3	4	5	6
Recall	81%	84%	90%	91%	91%	87%
Precision	100%	89%	86%	95%	88%	95%
AUC	91%	84%	81%	90%	77%	86%
F1	0.90	0.86	0.89	0.93	0.89	0.91

Table 4.2 Release of abductees: predictive results for our 2023 predictive reports

Release of abductees						
Time period	1	2	3	4	5	6
Recall	60%	92%	69%	71%	53%	73%
Precision	67%	92%	100%	100%	89%	92%
AUC	73%	93%	85%	86%	74%	83%
F1	0.63	0.92	0.82	0.83	0.67	0.82

words, we predicted, at the beginning of January 2023 (using data till the end of December 2022), the attacks by JNIM sometime in the next month, the next 2 months, and so forth upto a 6-month lead time window. We then did the same thing at the beginning of February 2023 using the data till the end of January 2023. We used 6 classifiers on our dataset: SVM, KNN, Random Forest, Gaussian Naïve Bayes, Multinomial Naïve Bayes, and Logistic Regression, together with the late fusion module that combines the results of these predictors. The model predicts whether kidnappings will occur or if abductees will be released, within a given timeframe. For example, if the offset is 2 then the model predicts whether or not abductees will be released anytime during the next 2 months. Tables 4.1 and 4.2 below depict the results of our predictions compared to the ground truth observed after the predictions we made.

References

AFP (2011). Two French nationals kidnapped in Mali. https://www.google.com/hostednews/afp
Al-Naba (2020). [Issue 240]. Islamic State, N/A.
Berger, F. (2023, March). *Kidnappings in Burkina Faso*. Globalinititive.net. https://globalinitiative.net/wp-content/uploads/2023/03/Flore-Berger-The-silent-threat-Kidnappings-in-Burkina-Faso-GI-TOC-March-2023.pdf
Cigar, N. (2009). Al-Qa'ida, the Tribes, and the Government: Lessons and Prospects for Iraq's Unstable Triangle (Strategic Studies Institute Monograph). U.S. Army War College. https://ssi.armywarcollege.edu/pubs/display.cfm?pubID=902
Nünlist, C. (2013, October). *Kidnapping for ransom as a source of terrorism funding*. ccs.ethz.ch. https://css.ethz.ch/content/dam/ethz/special-interest/gess/cis/center-for-securities-studies/pdfs/CSS-Analysis-141-EN.pdf
Reuters (2021). *French journalist kidnapped in northern Mali appears in video*. https://www.reuters.com/world/europe/french-journalistkidnapped-northern-mali-appears-video-2021-05-05/

Chapter 5
Attacks on and Targeting of Public Sites

Insurgents often deliberately target public sites for strategic reasons. Assuming that many, if not most are done with the intent to generate a psychological impact beyond the immediate victim, this often puts these kinds of attacks in the realm of terrorism. To be specific, attacking public sites allows insurgents to instill fear and disrupt societal order, ultimately undermining the public's confidence in government institutions. Targeting public sites serves as a means for insurgents to challenge the legitimacy of the state, demonstrating their ability to operate with impunity and undermine the government's authority. Public spaces such as markets, transportation hubs, prestigious hotels, popular restaurants and government buildings are often crowded and well known, making them attractive targets for inflicting mass casualties and generating widespread publicity. Symbolically, attacking public sites associated with the adversary's political, ideological or cultural identity can also convey messages of defiance, resistance, and ideological supremacy to both domestic and international audiences. By striking at the heart of civilian life, insurgents seek to destabilize communities, sow discord, and erode public trust in the ability of authorities to provide security and maintain order. Terrorism is "the use of violence or the threat of violence with the primary purpose of generating an effect beyond the immediate victims or objective, with a political motive" (Richards, 2014). Typically, deliberately targeting public sites is associated with the later part of phase 1 in Mao's model, after the organizational structure in an area of operations has been established.

JNIM has conducted several high-profile attacks on symbolic targets such as public places regularly visited by Westerners. A clear example is the attack on the Radisson Blu Hotel in Bamako, Mali by the Macina Liberation Front in which 170 individuals were held hostage and 20 were killed (BBC News, 2015). These attacks generally received much media attention, which JNIM exploited by making public statements linking the attacks to their overarching strategy. Rather than JNIM itself, its constituent units have a reputation for conducting terrorist attacks prior to becoming part of JNIM. In particular, al-Mourabitoun launched several high-profile attacks that generated publicity (Filiu, 2017). All these attacks can be appropriately labeled terrorist attacks, being politically motivated attacks that intend to generate a psychological impact beyond the immediate victims. According to the perpetrators' media releases, the attacks had classic short-term terrorist goals like gaining publicity to mark specific events or to install vengeance. However, the number of high-profile terrorist attacks with indiscriminate violence against civilians seems to have dropped after JNIM's creation.

It can be argued that after the French intervention in 2013, JNIM's predecessors used terrorism as an asymmetric tactic of low-level violence while the insurgency was in Mao's doctrinal phase of strategic defense. Its purpose then was to spread disorder, gain publicity, and occasionally install vengeance, thus giving the different insurgent armed groups the time and space to concentrate on survival and political organization. The visible intensification of guerrilla warfare since then, leading to the creation of JNIM and the subsequent escalation and expansion of violence on a regional scale, could be a sign that the insurgency is moving forward into the doctrinal phase of strategic stalemate. The terrorist attacks against high-profile targets seem to have ceased once they no longer supported JNIM's strategic goals. This led to an increase in attacks on less prominent public sites, intended to harm individuals with a political, strategic or military role. This information, however, is not encapsulated by our TP-rules.[1] The following instances of attacks on public sites further illustrate the evolution in JNIMs' strategies and tactics concerning attacks on public sites.

- **7 August 2015**: Attackers stormed the Byblos hotel in what military sources and local residents said appeared to be an attempt to kidnap Western hotel guests (France 24, 2015)
- **20 November 2015**: Twenty-one people were killed in an attack on the Raddison Blu Hotel in Bamako, Mali (CNN, 2015)
- **16 January 2016**: 28 people were killed and a further 56 injured after Islamist militants attacked a hotel in the capital, Ouagadougou, popular with foreigners (BBC, 2016)
- **14 August 2017**: At least 18 people have been killed in an attack by Islamist militants at a restaurant popular with foreigners in the capital of Burkina Faso (The Guardian, 2017)
- **12 January 2019**: Twelve civilians were killed on Thursday during a jihadist attack in the north of Burkina Faso (The Independent, 2019)
- **11 May 2020**: At least 20 civilians in the West African nation of Niger were killed in weekend attacks on villages near the border with Mali (BBC, 2020)

---

[1] The current codebook does not contain a variable capturing the 'prominence' of a public site. The codebook variable Act_Z_Struct_Publicsite covers public sites in general and is coded as a binary variable, i.e., it does not distinguish between an international hotel that hosts Westerners from a local market.

In the period between 2011 and 2022, the 2017 merger of several terrorist groups to form JNIM led to an increase in attacks on, and targeting of, public sites. Until 2017, JNIMs carried out attacks on public sites in 5 out of 84 months. Since 2017, this number has risen to 42 out of 72 months. Figure 5.1 visually represents the frequency of attacks on public sites during the studied timeframe and illustrates the upward trend in attacks on public sites. Additionally, this figure depicts the months when JNIM targeted public sites. The difference between attacking public sites and targeting them, is that targeting public sites does not have to involve an attack involving casualties, whereas attacking public sites does. Figure 5.2 presents the number of casualties resulting from attacks at public sites and the instances when public sites were targeted by JNIM. While the number of months with attacks on and targeting of public sites seems to increase (Fig. 5.1), the number of instances when a public site is targeted does not seem to increase significantly. This may mean that since the amalgamation of different groups in JNIM, rather than attacking public sites less frequently, the group is spreading this out more throughout the year. This may indicate a change in strategy or decision-making. In addition, Fig. 5.2 shows another remarkable fact. The number of instances involving targeting public places does not initially appear to correspond to the number of casualties resulting from these attacks. The spikes in casualties shown by the graph in 2021 and 2022, coupled with the relatively flat levels of targeting public sites, suggest an increase in effectiveness in attacking public sites over the years. This increase in effectiveness could also be interpreted as an increase in high-profile attacks, corresponding with the later phase 1 of Mao.

The increase in attacks on public sites shows how Western attempts to reform governance in the Sahel region have not prevented jihadist expansions. In fact, some

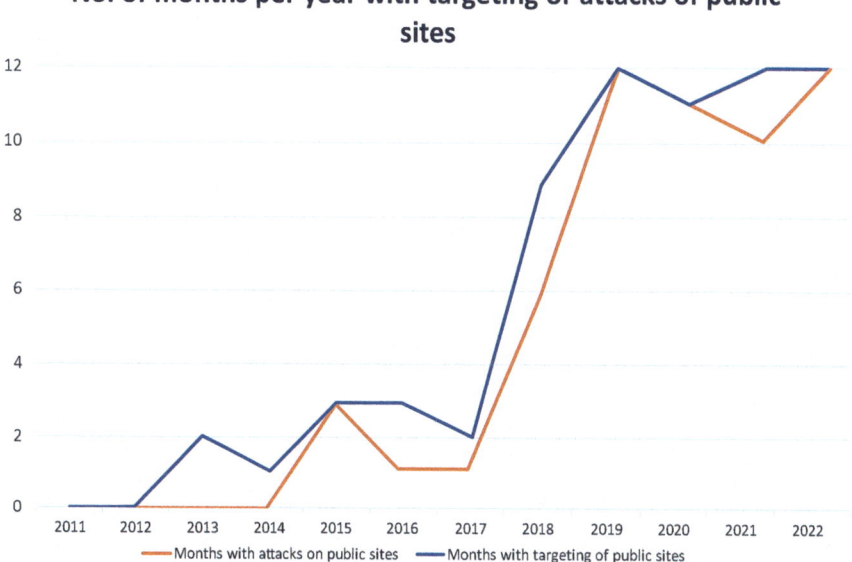

Fig. 5.1 Number of months per year when public sites were attacked or targeted

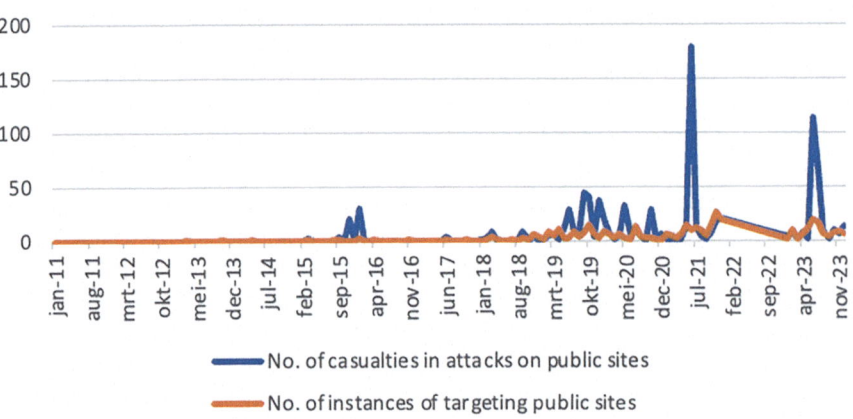

Fig. 5.2 Number of casualties connected to and instances of targeting public sites per month

argue they might even fuel them (CSIS, 2020). In addition, a UN report shows how the entry of the Wagner Group has led to a dramatic increase in political violence, with civilians increasingly becoming collateral damage (United Nations, 2023) or even being deliberately targeted as per the authoritarian approach to counterinsurgency (Routledge/Zhukov). The U.N. report also highlighted JNIM's ongoing focus on attacking Malian military forces within urban areas, which may also lead to an increase in the targeting of public sites. It noted that in the recent past, JNIM had enforced stringent societal and political rules on the residents of various locations where it holds influence. Additionally, the report indicated that there had been a rise in retaliatory actions against communities believed to have collaborated with the Malian junta forces (United Nations, 2023). Attacks on public sites are therefore carried out not so much to cause terror among the local population, but rather to target security forces and anyone who cooperates with them, or to send a political message.

Attacks on and targeting of public sites do not only have an impact on civilians with a military, political or strategic role, but also affect communities. This chapter's focus is on TP-Rules we derived to predict when JNIM targets or will commit attacks on public places. We will now briefly discuss the factors that played a significant role in the TP-rules that we derived. The variables most used for developing TP-Rules which predict attacks on public sites sometime in the next 4 months can be seen in Fig. 7.2. These variables, as well as others used in this chapter, are briefly explained below.

- *A travel ban was placed on Mali, Niger, or Burkina Faso.* By examining months when a travel ban was placed by foreign countries on Mali, Niger or Burkina

5 Attacks on and Targeting of Public Sites

Faso, we were able to derive TP-rules to predict attacks on public places sometime in the next 4 months.

- *A country or international organization freezes the assets of (members of) JNIM.* In months where a country or international organization froze the assets of (members of) JNIM, we were able to derive TP-rules for the next 4 month period.

We were able to derive TP-rules from the previously mentioned variables often in conjunction with other conditions. These conditions are briefly explained below.

- *JNIM was not in direct negotiations with the nation with which it is in conflict.* We were able to derive TP-rules predicting attacks on public places sometime in the next 5 months following months in which JNIM was not in direct negotiations with state governments and foreign states or international institutions froze asset(s) of individual member(s) of JNIM.
- *The leadership of JNIM was not elected.* In months when the leadership of JNIM was not elected, and the national government in the JNIM-operated region implemented a travel ban, we successfully formulated TP-rules that forecasted attacks on public places with sometime during the next 5 months.
- *JNIM did not communicate through print media (magazines, journals for publishing articles or advertisements in other publications).* In months when JNIM refrained from communication through print media, and the national government in the JNIM-operated region imposed a travel ban, we successfully formulated TP-rules that predicted attacks on public places sometime in the next 3 months.
- *The top leaders of JNIM were not under arrest or imprisoned in the coded period.* In months when the top leaders of JNIM were not under arrest or imprisoned, and there were asset freezes by international institutions on individual members of JNIM, we successfully formulated TP-rules predicting attacks on public places sometime in the next 4 months.
- *The top leaders of JNIM did not die in the period being coded.* During months when the top leaders of JNIM did not die, and the national government in the JNIM-operated region enforced a travel ban, we effectively devised TP-rules for forecasting attacks on public places sometime in the next 3 months.
- *The government did not warrant arrests of members of JNIM.* In months when the government refrained from issuing warrants for the arrests of JNIM members and there were asset freezes by international institutions on individual members of JNIM, we successfully formulated TP-rules for predicting attacks on public places sometime during the next 3 months.
- *JNIM did not communicate a message of solidarity.* During months when JNIM refrained from communicating a message of solidarity and the national government in the JNIM-operated region imposed a travel ban, we successfully formulated TP-rules for predicting attacks on public places sometime during the next 6 month window.
- *The national government did not raid JNIM facilities and locations.* In months when the national government refrained from raiding JNIM facilities and locations, and the government in the JNIM-operated region enforced a travel ban, we

effectively devised TP-rules for forecasting attacks on public places sometime during the next 6 months.

5.1 Attacks on Public Sites and Travel Bans

In months when a travel ban was imposed on Mali we established a TP-rule to anticipate attacks on public sites sometime during the next 4 months. As an illustration, in October 2018, Mali imposed a travel ban. Four months thereafter, on January 1, 2019, presumed Katiba Macina militants attacked a minibus between Bankass and Bandiagara. The vehicle was transporting livestock merchants involved in trading livestock rustled by Dozos, resulting in the death of the driver and injuries to two others.

5.1.1 TP-Rule PS-1

Attacks on public sites occur in months in which:

- Up to 4 months earlier, a travel ban was placed on the country where JNIM operates.

  Support = 0,28
  Probability = 81%, *Inverse Probability* = 70%, *Negative Probability* = 18%, *Lift* = 2.03

In total, we have derived TP-rules connecting this variable to seven other conditions with an offset ranging from the next 3 to 6 months. These TP-rules will be further explained below.

5.2 Attacks on Public Sites when Assets Are Frozen

During a period when international institutions or foreign states freeze assets of (individual members of) JNIM, we were able to derive a TP-rule predicting attacks on public sites sometime during the next 4 months. This scenario occurred in January 2020, and 4 months later, presumed Katiba Macina (JNIM) militants attacked the Dogon village of Tama (Bankass, Mopti), reportedly resulting in the death of 15 people.

5.2.1 TP-Rule PS-2

Attacks on public sites occur in months in which:

- Up to 4 months earlier, foreign states or international institutions froze asset(s) of individual member(s) of JNIM.

    Support = 0,28
    Probability = 80%, *Inverse Probability* = 70%, *Negative Probability* = 19%, *Lift* = 1.99

5.3 Attacks on Public Sites When Assets Are Frozen and the Group Was Not in Direct Negotiations

Similar to PS-2, TP-Rule PS-3 looks at months when assets of (individual members of) JNIM are frozen. However, PS-3 also incorporates an additional variable. In months when the assets of individual members of JNIM are frozen, and JNIM is not engaged in direct negotiations, we formulated a TP-rule to predict attacks on public places sometime in the next 5 months. These conditions were evident, for instance, in May 2020. Five months later, on August 3, 2020, presumed JNIM militants attacked the village of Koromatintin (Mali), resulting in the death of a woman and the seizure of livestock.

5.3.1 TP-Rule PS-3

Attacks on public sites occur in months in which:

- Up to 5 months earlier, foreign states or international institutions froze asset(s) of individual member(s) of JNIM
- Up to 5 months earlier, JNIM was not in direct negotiations with the nation with which it is in conflict.

    Support = 0,28
    Probability = 81%, *Inverse Probability* = 69%, *Negative Probability* = 19%, *Lift* = 2.02

5.4 Attacks on Public Sites When There Is a Travel Ban and the Leadership of the Group Is Not Elected

5.4.1 TP-Rule PS-4

Attacks on public sites occur in months in which:

- Up to 3 months earlier, a travel ban was placed on the country where JNIM operates.
- Up to 3 months earlier, the leadership of JNIM was not elected.

Support = 0,27
Probability = 81%, *Inverse Probability* = 68%, *Negative Probability* = 19%, *Lift* = 2.04

5.5 Attacks on Public Sites When There Is a Travel Ban and the Group Did Not Communicate through Print Media

Looking at TP-rule PS-5, attacks on public sites also occur during the 5 months after there was a travel ban instituted and JNIM did not communicate through print media (as far as we have been able to identify). To be able to derive this rule, we have looked at different ways JNIM communicates between the time period of 2011 up until 2022. In doing so, we also looked specifically at print media. This implies that the methods which JNIM uses to communicate might also reveal information about attacks on public places, albeit in combination with a travel ban.

5.5.1 TP-Rule PS-5

Attacks on public sites occur in months in which:

- Up to 5 months earlier, a travel ban was placed on the country where JNIM operates.
- Up to 5 months earlier, JNIM did not communicate through print media (magazines, journals for publishing in other publications).

Support = 0,27
Probability = 81%, *Inverse Probability* = 68%, *Negative Probability* = 19%, *Lift* = 2.04

5.6 Attacks on Public Sites when their Assets Are Frozen and the Top Leaders of JNIM Are Not under Arrest or Imprisoned

TP-rule PS-6 was derived using variables which looked at both occurrences when international institutions froze asset(s) of individual member(s) of JNIM and occurrences when top leaders of JNIM were not under arrest or imprisoned. By applying rule PS-6, we successfully forecasted attacks on public sites occurring during the 4 months after the freezing of JNIM assets and when leaders were not in custody. To illustrate, in July 2018, multiple states froze JNIM's assets following UN Security Council Guidelines, and none of the JNIM leaders were arrested or imprisoned. Subsequently, on November 11, 2018, militants from Katiba Macina attacked the village of Mamadaga, engaging in a confrontation with Dozos based in the village. The attack resulted in five casualties, and the assailants managed to escape.

5.6.1 TP-Rule PS-6

Attacks on public sites occur in months in which:

- Up to 4 months earlier, foreign states or international institutions froze asset(s) of individual member(s) of JNIM.
- Up to 4 months earlier, the top leaders of JNIM were not under arrest or imprisoned

  Support = 0,28
  Probability = 81%, *Inverse Probability* = 70%, *Negative Probability* = 18%, *Lift* = 2.03

5.7 Attacks on Public Sites when there Is a Travel Ban and the Top Leaders of the Group Did Not Die

In addition to TP-rule PS-6, we were able to derive a TP-rule using another variable about top leaders of JNIM in combination with a travel ban placed on the state where JNIM operates. When there is a travel ban placed on states where JNIM operates and top leaders of the group did not die, we were able to use TP-rules we derived to predict attacks on public sites sometime in the next three-month period.

5.7.1 TP-Rule PS-7

Attacks on public sites occur in months in which:

- Up to 3 months earlier, a travel ban was placed on the country where JNIM operates.
- Up to 3 months earlier, the top leaders of JNIM did not die.

Support = 0,24
Probability = 85%, *Inverse Probability* = 61%, *Negative Probability* = 22%,
Lift = 2.14

5.8 Attacks on Public Sites when Foreign States or Institutions Froze Assets of JNIM and the Government Did Not Warrant Arrests for Members of the Group

TP-rule PS-8, like several other rules, considers situations when foreign states and/or international institutions freeze assets of (individual members of) JNIM as one of its pre-conditions. The other condition that needs to be satisfied is: the national government of the country where JNIM operates did not issue arrest warrant for members of JNIM. If both these two conditions are met, then we can use PS-8 to predict attacks on public sites sometime in the next four-month time window. One example of this rule can be seen in February 2021, when these two conditions were satisfied. Four months later, on 10 July 2021, presumed Katiba Macina (JNIM) militants attacked the village of Yaro-Bombo (Mali). The militants killed six people including five men and one woman and burned the village. The results of PS-8 are displayed below.

5.8.1 TP-Rule PS-8

Attacks on public sites occur in months in which:

- Up to 4 months earlier, foreign states or international institutions froze asset(s) of individual member(s) of JNIM.
- Up to 4 months earlier, the government did not warrant arrests of members of JNIM.

Support = 0,28
Probability = 81%, *Inverse Probability* = 70%, *Negative Probability* = 18%,
Lift = 2.03

5.9 Attacks on Public Sites When a Travel Ban Is Placed on the State Were JNIM Operates and the Group Did Not Communicate a Message of Solidarity

We were also able to analyze the public communications and messages disseminated by JNIM from 2011 to 2022. In particular, when JNIM did not convey a message of unity while simultaneously a travel restriction was imposed on the region where it operates, we identified a TP-rule that allowed us to predict attacks on public sites sometime during the subsequent 6 months. This insight was gained through a focused examination of instances where the absence of solidarity messages coincided with travel bans within the operational area of JNIM. These conditions were met in March 2020, when JNIM did not convey any messages of solidarity and a travel ban was placed on Mali. Six months later, on 7 September 2020, presumed JNIM militants assaulted and wounded two elderly women and prevented others from entering a market in Petegoli.

5.9.1 TP-Rule PS-9

Attacks on public sites occur in months in which:

- 6 months earlier, a travel ban was placed on the country where JNIM operates.
- 6 months earlier, JNIM did not communicate a message of solidarity.

Support = 0,25
Probability = 80%, *Inverse Probability* = 63%, *Negative Probability* = 22%, *Lift* = 1.96

5.10 Attacks on Public Sites When a Travel Ban Is Placed on the State Were JNIM Operates and the Government Did Not Raid Facilities and Locations of the Group

Lastly, we were also able to derive a TP-rule using the travel ban variable and one other variable: occurrences when security forces did not raid JNIM facilities or locations. If these two preconditions were met, we were able to predict attacks on public sites sometime during the succeeding 6 month period. This rule, TP-rule PS-10, is depicted below.

5.10.1 TP-Rule PS-10

Attacks on public sites occur in months in which:

- Up to 6 months earlier, a travel ban was placed on the country where JNIM operates.
- Up to 6 months earlier, the government did not raid JNIM facilities and locations.

Support = 0,28
Probability = 83%, *Inverse Probability* = 70%, *Negative Probability* = 19%, *Lift* = 2.04

5.11 Conclusions

Over the past years, JNIM has been attacking public sites for various reasons. These attacks have been responsible for over hundreds of deaths across Mali, Niger and Burkina Faso between 2011 and 2022. In our research, we have found several conditions that can be linked to the attacks JNIM has been conducting across the Sahel. The following paragraphs describe some of these conditions. It should be noted that these relationships are not causative.

- When a travel ban is imposed on the state where JNIM operates (Mali, Niger, or Burkina Faso), attacks on public sites tend to follow sometime in the next 4 months.
- When the assets of individual members of JNIM are frozen by foreign states or international institutions, attacks on public sites tend to occur within the next 4 months.
- When both assets are frozen, and JNIM is not engaged in direct negotiations with the national government, attacks on public sites tend to occur within the subsequent 5 months.
- When a travel ban is imposed on the state where JNIM operates, and the leadership of JNIM is not democratically chosen, attacks on public places may occur within the next 3 months.
- When JNIM does not communicate via print media for a whole month, and a travel ban is imposed on Mali, Niger, or Burkina Faso, attacks on public sites tend to occur within the next 3 months.
- When the top leaders of JNIM are not under arrest or imprisoned, and assets are being frozen for individual members of JNIM, attacks on public sites occur during the subsequent 4 month time window.
- When top leaders of JNIM did not die in a month, and a travel ban was placed on Mali, Niger, or Burkina Faso, attacks on public sites tend to occur sometime during the following 3 months.

5.12 Predictive Model/Reports Results

- Attacks on public sites also seem to occur within the 4 months after months when the national government of the state where JNIM operates did not issue any warrants for members of JNIM, and assets are being frozen.
- When JNIM does not communicate any messages of solidarity for a whole month, and a travel ban is imposed on the state where JNIM operates, attacks on public places seem to occur sometime in the next 6 months.
- Lastly, when security forces of the national government of the state where JNIM operates do not raid any facilities or locations of JNIM, and there is a travel ban, attacks on public sites seem to occur during the following 6 months.

While these correlations do not imply causation, they offer valuable insights for shaping policy options or decisions in the fight against JNIM. Understanding when attacks on public sites are less likely to occur allows for more strategic resource allocation, enhancing the ability to counter various actions undertaken by JNIM. This information can guide decision-makers in devising effective measures to mitigate the impact of JNIM activities.

5.12 Predictive Model/Reports Results

Table 5.1 below shows the performance of our Northwestern Terror Early Warning System's (NTEWS) predictive model. We learned the predictive model using data up to the end of 2022 and then made monthly predictions during 2023. In other words, we predicted, at the beginning of January 2023 (using data till the end of December 2022), the attacks by JNIM sometime in the next month, the next 2 months, and so forth up to a 6-month lead time window. We then did the same thing at the beginning of February 2023 using the data till the end of January 2023. We used six classifiers on our dataset: SVM, KNN, Random Forest, Gaussian Naïve Bayes, Multinomial Naïve Bayes, and Logistic Regression, together with the late fusion module that combines the results of these predictors. The model predicts whether public sites will, or will not be targeted, within a given timeframe. For example, if the offset is 2 then the model predicts whether or not public sites will be targeted anytime during the next 2 months. Table 5.1 below depicts the results of our predictions compared to the ground truth observed after the predictions we made.

Table 5.1 Targeting of public sites results for our 2023 predictive reports

Targeting public sites						
Time period	1	2	3	4	5	6
Recall	81%	89%	86%	96%	96%	100%
Precision	87%	94%	95%	85%	85%	92%
AUC	84%	91%	88%	73%	69%	83%
F1	0.84	0.91	0.9	0.9	0.9	0.96

References

BBC (2016) Burkina Faso attack: Foreigners killed at luxury hotel. https://www.bbc.com/news/world-africa-35332792

BBC (2015). Mali hotel attack in Bamako: Two held. https://www.bbc.com/news/world-africa-34936420

BBC (2020). At least 20 civilians in the West African nation of Niger were killed in weekend attacks on villages near the border with Mali. https://www.bbc.com/news/world-africa-52612247

CSIS (2020). The Sahel summit: A conversation with experts on security, development, and governance. https://www.csis.org/events/sahel-summit-conversation-experts-security-development-and-governance

CNN (2015). Deadly Mali hotel attack: They were shooting at anything that moved. https://www.cnn.com/2015/11/20/africa/malishooting/index.html

Filiu, J.-P. (2017). Al-Qaida in the Islamic Maghreb and the dilemmas of jihadi loyalty. Perspectives on Terrorism, 11(6), 3–16. https://pt.icct.nl/article/al-qaida-islamic-maghreb-and-dilemmas-jihadi-loyalty

France 24 (2015). Attackers stormed the Byblos hotel in what military sources and local residents said appeared to be an attempt to kidnap Western hotel guests. https://www.france24.com/en/20150807-mali-deadly-hostage-seige-hotel-sevare

Richards, A. (2014). Conceptualizing terrorism. *Studies in Conflict & Terrorism, 37*(3), 213–236. https://doi.org/10.1080/1057610X.2014.872023

The Guardian (2017). Burkina Faso: at least 18 dead in restaurant attack. https://www.theguardian.com/world/2017/aug/14/terror-attackrestaurant-burkina-faso-many-dead

The Independent (2019). Twelve civilians were killed on Thursday during a jihadist attack in the north of Burkina Faso. https://www.independent.co.uk/news/world/africa/burkina-faso-jihadist-attack-civilians-killed-north-a8728961.html

United Nations (2023). Mali: Over 500 people killed in anti-terrorist operation in Moura in March 2022 – UN report. https://www.ohchr.org/en/press-releases/2023/05/mali-over-500-people-killed-anti-terrorist-operation-moura-march-2022-un

Chapter 6
Targeting of Security Professionals or Security Installations

Insurgents frequently target security personnel and installations for strategic, practical, and symbolic reasons. Security personnel symbolically represent the frontline defense of the state and are tasked with maintaining law and order, protecting civilians, and thwarting insurgent activities. By attacking security personnel, insurgents seek to undermine the state's authority, disrupt its ability to enforce security, and weaken public confidence in the government's ability to provide protection. Security installations such as military bases, police stations, and checkpoints are targeted for their strategic significance, as they are often strategically located to control territory or important infrastructure. They are targeted for practical reasons as they often contain valuable equipment, supplies, intelligence, and personnel crucial to the state's security apparatus. Additionally, attacking security installations allows insurgents to project power, intimidate security forces, and demonstrate their operational capabilities.

It was already mentioned that JNIM uses a combination of terrorism and guerrilla tactics. Its use of terrorism against high-profile targets and the indiscriminate use of violence against civilians seems to have stopped after the announcement of its creation in 2017. In contrast, the number of attacks against both state and international security forces has dramatically increased (Eizenga and Williams, 2019). Thus, the 2017 merger of Ansar Dine, AQIM Sahara, Al Mourabitoun, and Katiba Macina into JNIM seems to be aimed at filling the security vacuum caused by increasingly weak governance in the Sahel, particularly in Mali. Initially, JNIM targeted security forces mostly with (improvised) explosive devices (IEDs). Currently, armed assaults, assassinations, and ambushes are increasingly common (Zimmerer, 2022, p. 504). As mentioned before, the organization promotes action against security forces rather than against civilians (UNSC, 2018; Pollicini, 2021, p. 1101; Zimmerer, 2022, p. 495). This has resulted in numerous attacks against the Malian Defence and Security Forces (MDSF), the Joint Force of the Group of Five for the Sahel (G5 Sahel), the United Nations Multidimensional Integrated

Stabilization Mission (MINUSMA), and French forces. The focus on security forces undermines the legitimacy of the Malian government and those who support it, but it also sabotages law enforcement (Pollicini, 2021, p. 1093). The latter effect provides an environment in which criminal enterprises can thrive. Furthermore, targeting security forces aligns with AQ's broader global effort to brand itself as more moderate than its competitors. Besides these guerrilla attacks on security forces, JNIM also conducted several high-profile attacks on symbolic targets, often followed by public statements linking the attacks to the group's overarching strategy. These attacks have generated much attention. However, despite its claims that they focus their targeting against security forces, JNIM's violent actions have increasingly caused civilian casualties, and this increase has coincided with the expansion of JNIM's activities to Burkina Faso. This increase might be a means of responding to competition with other non-state armed groups, as well as a means of creating new front lines.

In the first 4 months of 2019, there was an average of 32 violent incidents attributed to JNIM in Burkina Faso, Mali, and Niger per month. This is in contrast to an average of 41 monthly incidents during the same period in 2020 and 59 in 2021. This represents a respective increase of 43% and 84% compared to the 2019 levels. Moreover, it should be emphasized that assaults are often directed specifically at UN forces. These UN troops have been engaged in the MINUSMA operation since 2013, working towards aiding Mali's transitional authorities in stabilizing the country and implementing the transitional roadmap. JNIMs' core message revolves around resisting foreign forces, which they label as invaders or occupiers. Furthermore, attacking them highlights their presence, assisting and reinforcing the narrative of the non-Muslim invaders and the local government being apostates.

Figure 6.1 illustrates the upward trend in targeting of security forces and installations in months per year. The targeting of security installations includes, for example, the targeting of barracks and police stations. It specifically involves the targeting of infrastructure, with possible casualties or deaths as a side effect. For the targeting of security personnel, all incidents not involving the targeting of security infrastructure were coded. Examples include the targeting of convoys or individuals with a security profession. Both definitions of the variables "Act_Z_Prof_Security_Force" and "Act_Z_Struc_Security_Installation" involve encoding not the number of casualties, but the number of times a targeting occurs.

Figure 6.2 illustrates the relationship between targeting security installations and professionals. From this graph, one can conclude that from 2017, JNIM started to target both security professionals and installations more often with their actions.

The following list further illustrates the change in JNIM's focus from attacking civilian targets towards attacking on security forces.

- **1 July 2018**: JNIM claims that Saad al Ansari, described as a local Malian, carried out a suicide bombing by driving his explosives-laden vehicle into a French patrol in the city of Gao. French military sources confirmed to France24 that a military patrol consisting of approximately 30 troops was indeed attacked in the city. The explosion resulted in at least eight French troops sustaining injuries, and tragically, two civilians lost their lives. (Long War Journal, 2018)

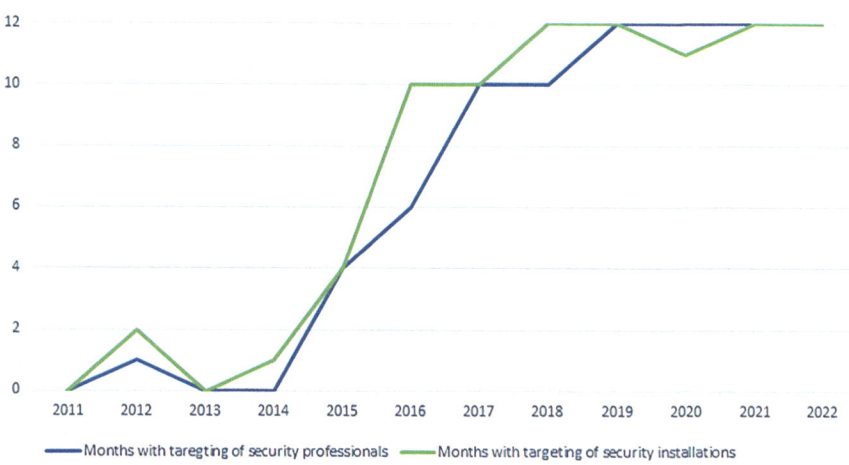

Fig. 6.1 illustrates the increased targeting of security forces or installations from 2013 until 2022

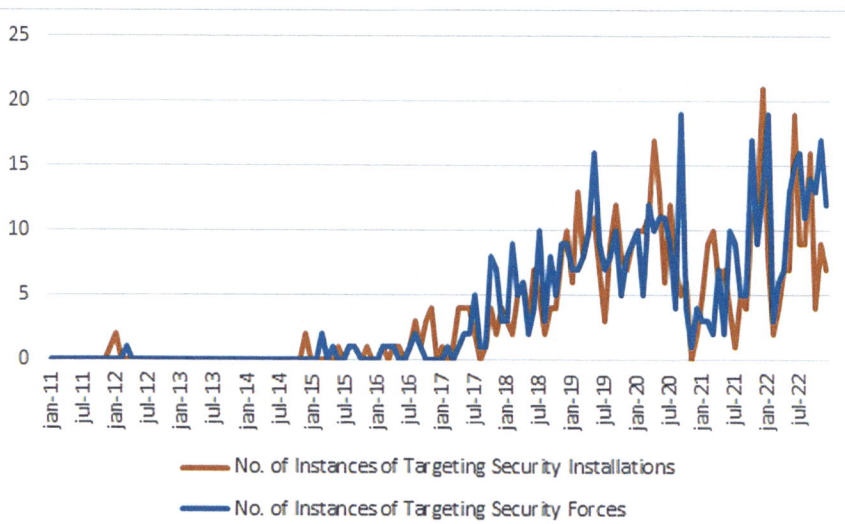

Fig. 6.2 Instances of targeting of security professionals or sites per month (2011-2022)

- **20 January 2019**: JNIM attacked a United Nations base in the northern Malian city of Aguelhok. According to the UN, at least 10 Chadian peacekeepers were killed and 25 others wounded (Weiss, 2019).
- **21 January 2019**: JNIM executed a suicide attack on a Malian military base in the northern town of Tarkint, leading to the death of two soldiers and injuries to ten others (Weiss, 2019).
- **10 February, 2019**: At least 40 people have been killed and 60 more are missing following an attack by extremists on two military camps in Mali. (Deutsche Welle, 2020)
- **17 March 2019**: According to a local mayor, armed individuals temporarily took control of an army base in Central Mali. The incident in the Mpoti region resulted in the loss of 16 Malian soldiers' lives, and the base was set ablaze, with indications suggesting the theft of weapons as well (Africa News, 2019).
- **23 July 2022**: The assault on the Kati military base near Bamako resulted in the loss of one soldier's life, and six additional individuals sustained injuries. (Al Jazeera, 2022).

Targeting of security forces and installations is part of the modus operandi of JNIM when it comes to conveying their political messages. This chapter focuses on TP-Rules developed to forecast instances when JNIM refrains from targeting security professionals or installations. The upcoming paragraphs provide a brief overview of the influential variables in the derivation of these TP-rules. Figure 6.2 displays the primary variables employed for constructing TP-Rules that predict the absence of targeting security professionals or installations sometime during the next 2 months. Subsequent to this, a concise explanation of these variables and others utilized in this chapter is presented below.

- *The national government's security forces did not execute civilians.* Through an analysis of months in which the security forces of the national government refrained from executing civilians, we formulated TP-rules capable of predicting the absence of targeting security professionals anytime during the next 2 months.
- *JNIM did not discuss its campaign such as whether or not it had been successful or had been achieving its objectives.* By investigating months during which the JNIM refrained from addressing the success or achievement of its campaign objectives, we developed TP-rules with the capability to predict the absence of targeted attacks on civilians during the next 2-month window.
- *A foreign state or international organization did not freeze the assets of (members of) JNIM.* By examining months in which a country or international organization refrained from freezing the assets of (members of) JNIM, we formulated TP-rules capable of predicting the non-targeting of security professionals or installations during the next 2 month time frame.

TP-rules were derived from the aforementioned variables, frequently in conjunction with other conditions. Additionally, TP-rules were formulated based on different variables, incorporating one, four, or five-month offsets. The details of these conditions are briefly explained below.

6.1 No Targeting of Security Forces When There Is No Travel Ban Imposed... 75

- *JNIM did not discuss their campaign in its communications.* By analyzing periods during which JNIM refrained from addressing their campaign in communications, and when a country or international organization did not impose asset freezes on (members of) JNIM, we developed TP-rules that can forecast the absence of targeting towards security professionals or installations during the next 1 month.
- *JNIM did not issue claims of responsibility for violent events.* By analyzing periods during which JNIM did not issue claims of responsibility for violent events, and when a country or international organization did not impose asset freezes on (members of) JNIM, we developed TP-rules that forecast the absence of targeting towards security professionals or installations during the next month.
- *The national government did not freeze assets of JNIM.* Through an examination of months in which the national government refrained from freezing assets of JNIM and JNIM avoided discussing its strategy in communications, we established TP-rules with a predictive capacity covering the next four-month time period.
- *JNIM did not discuss its campaign such as whether or not it had been successful or had been achieving its objectives.* By analyzing periods during which JNIM did not discuss its campaign such as whether or not it had been successful or had been achieving its objectives, and when the national government did not declare a state of emergency, we developed TP-rules that can forecast the absence of targeting towards security professionals or installations sometime during the next 5 months.

6.1 No Targeting of Security Forces When There Is No Travel Ban Imposed on the Country Where JNIM Operates and JNIM Did Not Discuss Their Strategy

In months where the national government where JNIM operates does not institute a travel ban and the group did not discuss their strategy in its communications, we were able to derive a TP-rule to predict the absence of targeting of security forces during the next 4 months.

6.1.1 TP-Rule P6-1

No targeting of security forces occurs in months in which:

- Up to 4 months earlier, the national government where JNIM operates, did not institute a travel ban.
- Up to 4 months earlier, JNIM did not discuss its strategy in its communications.

Support = 0,34
Probability = 73%, *Inverse Probability* = 80%, *Negative Probability* = 16%,
Lift = 1.74

6.2 No Targeting of Security Forces When Foreign States or International Institutions Did Not Freeze Assets and JNIM Did Not Discuss Their Campaign

Besides JNIM not discussing strategy, not discussing their campaign also seems to influence the absence of targeting of security forces. Combined with the absence of sanctions leading to freezing of assets of (individual members of) JNIM, we were able to predict the absence of targeting of security forces during the next one-month time frame. Both conditions were met in November 2016, leading to no occurrences of targeting of security forces in December 2016. The TP-rule describing these results is depicted below.

6.2.1 TP-Rule P6-2

No targeting of security forces occurs in months in which:

- Up to 1 month earlier, the national government did not freeze assets of JNIM.
- Up to 1 month earlier, JNIM did not discuss their campaign in its communications.

Support = 0,32
Probability = 81%, *Inverse Probability* = 74%, *Negative Probability* = 19%, *Lift* = 1.86

6.3 No Targeting of Security Forces When Foreign States or International Institutions Did Not Freeze Assets and JNIM Did Not Issue Claims of Responsibility

As indicated in the earlier descriptions of rules, the content of messages disseminated by JNIM plays a crucial role in predicting the decision not to target security forces. This is mirrored in TP-rule P6-3. By applying this TP-rule, we successfully predicted the absence of targeting security forces during the next 1 month. The rule comes into effect when two conditions are satisfied. Firstly, foreign states or international institutions did not freeze the assets of JNIM, and secondly, JNIM did not claim responsibility for violent events such as attacks or abductions.

6.3.1 TP-Rule P6-3

No targeting of security forces occurs in months in which:

- Up to 1 month earlier, foreign states or international institutions did not freeze assets of (individual members of) JNIM.
- Up to 1 month earlier, JNIM did not issue claims of responsibility for violent events.

  Support = 0,31
  Probability = 81%, *Inverse Probability* = 71%, *Negative Probability* = 20%,
  Lift = 1.88

A rule for predicting the absence of targeting of security forces with the same conditions as TP-rule P6-3 was also derived with an offset of 2 months, as depicted below.

6.3.2 TP-Rule P6-4

No targeting of security forces occurs in months in which:

- Up to 2 months earlier, foreign states or international institutions did not freeze assets of (individual members of) JNIM.
- Up to 2 months earlier, JNIM did not issue claims of responsibility for violent events.

  Support = 0,28
  Probability = 78%, *Inverse Probability* = 62%, *Negative Probability* = 25%,
  Lift = 1.81

6.4 No Targeting of Security Forces When the National Government Security Forces Did Not Execute Civilians and the Government Did Not Declare a State of Emergency

While the prior rules focused on conditions related to JNIM's behavior, the next rule specifically pertains to the actions of the national government or security forces. By examining the occurrences of security forces executing civilians and the government declaring a state of emergency, we successfully predicted the absence of targeting security forces during the next 2 months. An illustration of TP-rule P6-4 occurred in January 2017 when both conditions were met, and 2 months later, there was no targeting of security forces.

6.4.1 TP-Rule P6-5

No targeting of security forces occurs in months in which:

- Up to 2 months earlier, the national government's security forces did not execute civilians.
- Up to 2 months earlier, the national government did not declare a state of emergency.

   Support = 0,31
   Probability = 81%, *Inverse Probability* = 72%, *Negative Probability* = 19%, *Lift* = 1.90

6.5 No Targeting of Security Forces When JNIM Did Not Discuss Its Campaign and the National Government Did Not Declare a State of Emergency

TP-Rule P-6-5 was derived to allow us to predict the absence of targeting of security forces 2 months in advance. For this rule to be viable, two variables must be satisfied. One variable requires that JNIM does not discuss its campaign in any of its messages. The second variable is satisfied if the national government of the state where JNIM operates does not declare a state of emergency. An example of this rule can be seen in April of 2017 when no security forces were targeted. Two months prior, all of the conditions in rule P6-5 had been met.

6.5.1 TP-Rule P6-6

No targeting of security forces occurs in months in which:

- Up to 2 months earlier, JNIM did not discuss its campaign such as whether or not it had been successful or had been achieving its objectives.
- Up to 2 months earlier, the national government did not declare a state of emergency.

   Support = 0,26
   Probability = 79%, *Inverse Probability* = 61%, *Negative Probability* = 25%, *Lift* = 1.83

A rule for predicting the absence of targeting of security forces with the same conditions as TP-rule P6-5 was also derived with an offset of 5 months, as depicted below.

6.5.2 TP-Rule P6-7

No targeting of security forces occurs in months in which:

- Up to 5 months earlier, JNIM did not discuss its campaign such as whether or not it had been successful or had been achieving its objectives.
- Up to 5 months earlier, the national government did not declare a state of emergency.

Support = 0,25
Probability = 74%, *Inverse Probability* = 64%, *Negative Probability* = 22%, *Lift* = 1.88

6.6 Conclusions

The consideration of security forces as targets by JNIM is a critical factor in formulating effective policy options to counter their violent attacks. In this chapter we have elaborated on seven rules that shed light on the circumstances surrounding JNIM's targeting decisions, particularly when they abstain from targeting security forces. Six of these rules are intricately connected to how JNIM communicates and the content of their messages. While the relationships portrayed in this chapter do not establish causation, they provide valuable insights, offering a clearer understanding of JNIM's operational patterns. The following paragraphs summarize some of the conditions associated with the non-targeting of security forces by JNIM.

- When JNIM refrains from discussing strategy in its communications up to 4 months in advance, and there is no travel ban imposed on the state where JNIM operates, targeting of security forces does not seem to occur.
- In months where there are no assets being frozen for individual members of JNIM, and JNIM does not discuss its campaign in their communications, targeting of security forces does not seem to occur during the following month.
- When JNIM avoids claiming responsibility and there are no assets being frozen for individual members of JNIM by foreign states or international institutions, security forces do not seem to be targeted by JNIM during the next 1 month period as well as the next 5 month period.
- In months when security forces do not execute civilians, and there is no state of emergency, targeting of security forces does not seem to occur within the next 2 months.
- Targeting of security forces does not seem to occur when up to two or up to 5 months in advance JNIM is not discussing its campaign in its communications, and the national government did not declare a state of emergency.

These relationships, although not establishing direct causation, offer substantial insights. The rules we have derived provide an improved understanding of JNIM's behavioral patterns, particularly in terms of their intentions to target security forces. Effectively mitigating or preventing such targeting can significantly enhance the operational efficiency of security forces. By offering a degree of foresight, the rules presented in this chapter serve as a valuable asset, empowering security forces to navigate the challenges of asymmetric conflict more effectively. In essence, they contribute to a proactive approach, allowing security forces to better allocate resources, refine strategies, and bolster their preparedness in the dynamic landscape of counterterrorism efforts.

6.7 Predictive Model/Reports Results

Table 6.1 below shows the performance of our Northwestern Terror Early Warning System's (NTEWS) predictive model. We learned the predictive model using data up to the end of 2022 and then made monthly predictions during 2023. In other words, we predicted, at the beginning of January 2023 (using data till the end of December 2022), the attacks by JNIM sometime in the next month, the next 2 months, and so forth up to a 6-month lead time window. We then did the same thing at the beginning of February 2023 using the data till the end of January 2023. We used 6 classifiers on our dataset: SVM, KNN, Random Forest, Gaussian Naïve Bayes, Multinomial Naïve Bayes, and Logistic Regression, together with the late fusion module that combines the results of these predictors. The model predicts whether security professionals or installations will, or will not be targeted, within a given timeframe. For example, if the offset is 2 then the model predicts whether or not security professionals or installations will be targeted anytime during the next 2 months. Table 6.1 below depicts the results of our predictions compared to the ground truth observed after the predictions we made.

Table 6.1 Targeting of security professionals results for our 2023 predictive reports

Targeting Security Forces						
Time period	1	2	3	4	5	6
Recall	95%	85%	95%	91%	100%	96%
Precision	100%	94%	91%	100%	85%	96%
AUC	97%	88%	88%	96%	75%	91%
F1	0.97	0.9	0.93	0.95	0.92	0.96

References

Africa News. (2019, March 17). Gunmen raid Mali military camp, 16 soldiers killed. Africanews. https://www.africanews.com/2019/03/17/gunmen-raid-mali-military-camp-16-soldiers-killed/

Al Jazeera (2022). Al-Qaeda affiliate claims deadly Mali attack. https://www.aljazeera.com/news/2022/7/23/alqaeda-affiliate-group-claims-deadly-mali-attack

Deutsche Welle (2020). Attack on Mali army kills dozens of soldiers, https://www.dw.com/en/attack-on-mali-armykills-dozens-of-soldiers/a-51306367

Eizenga D, Williams W (2019). The Puzzle of JNIM and Militant Islamist Groups in the Sahel – Africa Center. In: Africa Center. https://africacenter.org/publication/puzzle-jnim-militant-islamist-groups-sahel/

JNIM claims suicide bombing on French troops in Gao | FDD's Long War Journal. (2018, July 2). https://www.longwarjournal.org/archives/2018/07/jnim-claims-suicide-bombing-on-french-troops-in-gao.php

Pollicini, L. (2021). A case of violent corruption: JNIM's insurgency in Mali (2017–2019). *Small Wars & Insurgencies, 32*(7), 1092–1116.

United Nations Security Council (2018). The Situation in Mali, S/2018/866.

Weiss C (2019). JNIM claims suicide assault on Malian military. In: FDD's Long War Journal. https://www.longwarjournal.org/archives/2019/01/jnim-claims-suicide-assault-on-malian-military.php

Zimmerer, M. (2022). Terror in West Africa: a threat assessment of the new Al Qaeda affiliate in Mali. *Critical Studies on Terrorism, 12*(3), 491–511.

Chapter 7
Targeting of Civilians

Insurgents often deliberately target civilians as part of their strategy for several reasons (Neumann & Smith, 2005). First and foremost, attacking civilians serves to instill fear and terrorize populations, creating a sense of insecurity and vulnerability among civilians. By targeting non-combatants, including women, children, and the elderly, insurgents seek to undermine societal stability, fracture community cohesion, and ultimately erode trust in government institutions. Because civilians are perceived as soft targets, often lacking the protection and defenses of military or security personnel, they are also more susceptible to attacks. Insurgents may also target civilians to incite sectarian or ethnic violence, exacerbate existing tensions, and deepen societal divisions. Symbolically, attacking civilians can serve as a means to garner attention, generate headlines, and attract sympathy or support from sympathetic groups or sympathizers.

Al-Zawahiri stated in his 'General guidelines for Jihad' that members and affiliates of AQ should "refrain from killing and fighting against non-combatant women and children, and even if they are families of those who are fighting against us, refrain from targeting them as much as possible" (Al-Zawahiri, 2013). In the same document, another guideline called to "refrain from targeting enemies in mosques, markets and gatherings where they mix with Muslims or with those who do not fight us" (Al-Zawahiri, 2013). In practice, however, AQ members do seem to deliberately target non-combatants. Although al-Muqrin's writing preceded the aforementioned guidelines of al-Zawahiri, he doesn't make any distinction between combatants and non-combatants. In fact, he specifically mentions certain categories of non-combatant Jews, Christians and apostates who should be targeted. Furthermore, he also specifically mentions the spreading of terror as a mandatory duty, "as required by the holy verse" (Cigar, 2009).

Terrorism is seen as a useful tactic within the context of orchestrating an insurgency (Jones, 2017). In the initial stages of an insurgency, effective tactics should (amongst others) exploit government weaknesses, undermine government legitimacy and provoke an overreaction from the government (Jones, 2017). When relying on the utility of terrorism within an insurgency strategy, the insurgents seek to subsequently disorient the population, trigger a response from the government, and gain legitimacy by being considered a political actor to be reckoned with (Jones, 2017). To do so requires the insurgents to gain strategic momentum, yet avoid the 'escalation trap', which means that a government reacts in such an escalatory way that they succeed in effectively destroying the insurgents' organization. Furthermore, to effectively utilize terrorism, the insurgents must have the capability and the will to escalate to unspeakable levels of violence, while simultaneously mitigating the risk of erosion of public sympathy (Neuman et al., 2005). In contrast, some argue that jihadists view violence and death as more than instrumental in the achievement of their aims: it is an object of commitment in itself (Clarke, 2019). This seems more in line with al-Muqrin's statement that the spread of terror is a holy duty (Cigar, 2009).

JNIM and/or its constituents prior to the merger have conducted numerous terrorist attacks that target civilians (Fig. 7.1). The context and details of these attacks require careful analysis to determine the exact intent of a particular attack against a specific target. As the following brief overview illustrates, the details of the attacks can differ widely.

- **January 2016:** Thirty people were killed in attacks on a restaurant and a hotel in Ouagadougou. (Cape Times, 2018)
- **11 January 2017:** Six people were killed in an attack on an armed Tuareg group between the towns of Goundam and Douekire near Timbuktu (BBC Monitoring, 2017)
- **12 January 2019:** Twelve civilians killed in an attack on a village market near Gasseliki (Agence France Presse, 2019)
- **7 November 2019:** Thirty-seven civilians were killed in an attack on a convoy transporting workers of a Canadian-based intermediate gold producer. (China Daily, 2019)
- **20 June 2022:** 132 civilians were killed during the night between 18 and 19 June near the localities of Dialassagou, Dianweli and Deguessagou. (CE Noticias Financieras, 2022)
- **14 October 2022:** Thirty-nine people were killed following the explosion of a bomb on a bus in the vicinity of the Malian town Tilé. (CE Noticias Financieras, 2022)

This chapter's focus is on TP-Rules we derived to predict when JNIM will target civilians. We will now briefly discuss the factors that played a significant role in the TP-rules that we derived. The variables most used for developing TP-Rules which predict attacks on civilians during the next four-month time frame can be seen in Fig. 7.2. These variables, as well as others used in this chapter, are briefly explained below.

- *A travel ban was placed on Mali, Niger, or Burkina Faso.* By examining months when a travel ban was placed by foreign countries on Mali, Niger or Burkina

7 Targeting of Civilians

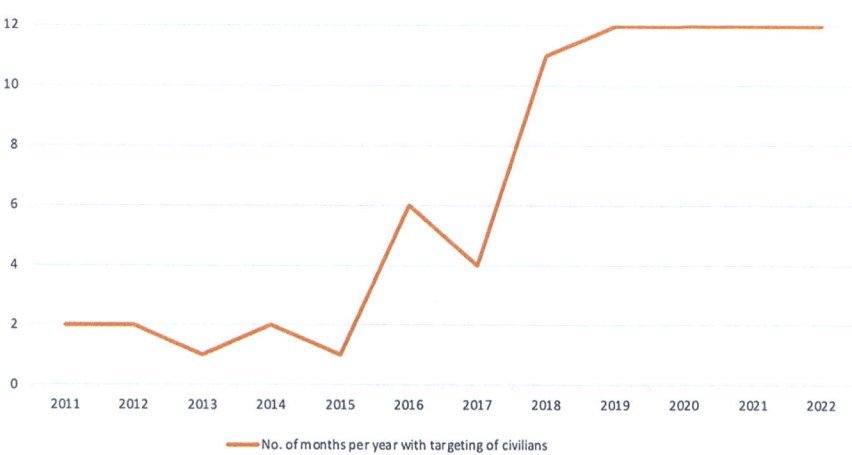

Fig. 7.1 No. of months per year with targeting of civilians

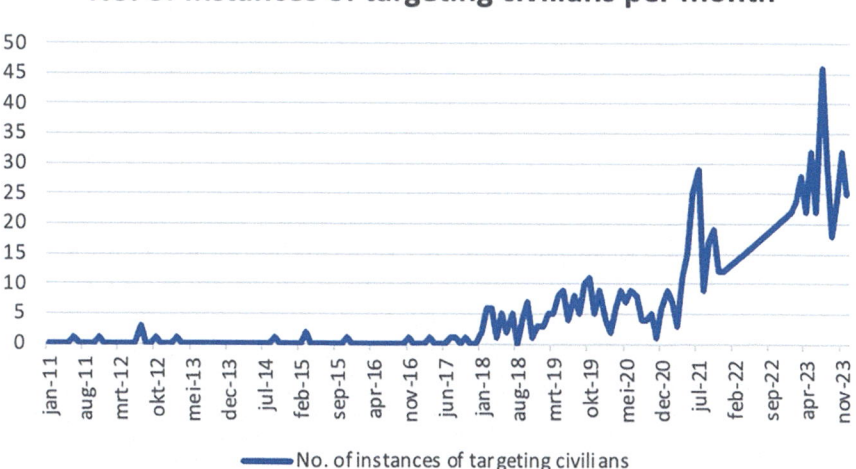

Fig. 7.2 Number of instances targeting of civilians

Faso, we were able to derive TP-rules to predict attacks on civilians during the subsequent 4 months.

- *A country or international organization freezes the assets of (members of) JNIM.* In months where a country or international organization froze the assets of (members of) JNIM, we were able to derive TP-rules predicting attacks targeting civilians sometime during the next 4 months.

We were able to derive TP-rules from the previously mentioned variables often in conjunction with other conditions. These conditions are briefly explained below.

- *The top leaders of JNIM are not imprisoned or under arrest.* We were able to derive TP-rules predicting attacks on civilians during the next 4 months after months in months in which the top leaders of JNIM were not imprisoned or under arrest and foreign states or international institutions froze asset(s) of individual member(s) of JNIM.
- *Members of JNIM are not under arrest.* In months when members of JNIM are not under arrest, and foreign states or international institutions froze asset(s) of individual member(s) of JNIM, we successfully formulated TP-rules that forecasted attacks on public places with sometime during the next 4 months.
- *The government did not raid JNIM facilities and locations.* In months when the government did not raid JNIM facilities and locations, and foreign states or international institutions froze asset(s) of individual member(s) of JNIM, we successfully formulated TP-rules that predicted attacks on public places within the next 4 months.
- *The government in the nation where JNIM operates received foreign military aid.* In months when the government in the nation where JNIM operates received foreign military aid, and there were asset freezes by international institutions on individual members of JNIM, we successfully formulated TP-rules predicting attacks on public places sometime in the next 4 months.

7.1 Targeting of Civilians When the National Government Instituted a Travel Ban

In months where the national government where JNIM operates institutes a travel ban, we were able to derive a TP-rule to predict JNIM attacking civilians for reasons that were not linked to their specific identity, sometime during the following 5 months. For example, the prerequisite condition was met in February 2020, when a travel ban was placed on Mali by several (neighboring) states. Five months later, in July 2020, JNIM militants abducted a man, assaulted tobacco vendors and burned their goods at a market in Kanioume, Mali. TP-rule TC-1 is depicted below.

7.1.1 TP-Rule TC-1

JNIM attacks on civilians for reasons not linked to their identity occur in months in which:

- Up to 5 months earlier, the national government where the JNIM operates, instituted a travel ban.

Support = 0,34
Probability = 98%, *Inverse Probability* = 65%, *Negative Probability* = 27%,
Lift = 1.9

7.2 Targeting of Civilians When Foreign States or International Institutions Freeze Assets

Besides states placing a travel ban on Mali, Burkina Faso or Niger, we have also found that the freezing of assets of members of JNIM is influencing the targeting of civilians for reasons not specifically linked to their identity. Using the occurrences of asset freezing of (individual) members of JNIM, we were able to predict the targeting of civilians sometime during the next four-month time frame.

7.2.1 TP-Rule TC-2

Attacks on civilians for reasons not linked to their identity occur in months in which:

- Up to 4 months earlier, foreign states or international institutions froze asset(s) of individual member(s) of the JNIM.

  Support = 0,34
  Probability = 96%, *Inverse Probability* = 65%, *Negative Probability* = 27%,
  Lift = 1.87

The freezing of assets of individual members of JNIM by foreign states or international institutions is a variable we have successfully linked to occurrences of civilian targeting in multiple ways, often in conjunction with an additional variable. When combined with factors such as the frequency of government raids on JNIM locations, we achieved a more accurate prediction of civilian targeting. The subsequent four paragraphs will provide a detailed description of these TP-rules.

7.3 Targeting of Civilians When Foreign States or International Institutions Freeze Assets and Top Leaders of JNIM Are Not Under Arrest

The anticipation of civilian targeting can be heightened when the freezing of assets is coupled with additional conditions. According to TP-rule TC-3, the absence of JNIM leaders being either arrested or imprisoned, in conjunction with asset freezing, serves as an indicator, allowing for the prediction of civilian targeting sometime during the following 4 months. This comprehensive approach provides a more nuanced understanding of the factors contributing to the potential threat against civilians, emphasizing the significance of combining multiple elements for more accurate predictions.

7.3.1 TP-Rule TC-3

Attacks on civilians for reasons not linked to their identity occur in months in which:

- Up to 4 months earlier, foreign states or international institutions froze asset(s) of individual member(s) of the JNIM.
- Up to 4 months earlier, the top leaders of the JNIM were not imprisoned or under arrest.

  Support = 0,34
  Probability = 98%, *Inverse Probability* = 65%, *Negative Probability* = 27%, *Lift* = 1.90

7.4 Targeting of Civilians When Foreign States or International Institutions Freeze Assets and Members of JNIM Are Not Under Arrest

Similar to TP-rule TC-3, TP-rule TC-4 focuses on the correlation between asset freezing and another variable. In this case, it relates to the status of JNIM members avoiding arrest. When members of JNIM remain unrestrained by arrests and at the same time foreign states or international institutions freeze assets belonging to these members, we were able to predict the targeting of civilians sometime during the next 4 months. Again, this comprehensive methodology offers a broader insight into the factors contributing to potential threats against civilians when assets of JNIM are being frozen. It underscores the importance of combining various conditions to enhance the precision of predictions and reinforce the understanding of the complex dynamics involved in such situations.

7.4.1 TP-Rule TC-4

Attacks on civilians for reasons not linked to their identity occur in months in which:

- Up to 4 months earlier, foreign states or international institutions froze asset(s) of individual member(s) of the JNIM.
- Up to 4 months earlier, members of the JNIM were not under arrest.

Support = 0,34
Probability = 98%, *Inverse Probability* = 65%, *Negative Probability* = 27%,
Lift = 1.90

7.5 Targeting of Civilians When Foreign States or International Institutions Freeze Assets and the National Government Did Not Raid JNIM Facilities or Locations

The third TP-rule, which assesses the targeting of civilians through the freezing of assets, incorporates the frequency of government raids on JNIM facilities or locations as a second variable. During months when foreign states or international institutions froze assets of individual JNIM members and the national government refrained from raiding JNIM facilities or locations, we successfully predicted the targeting of civilians sometime during the succeeding 4 month time period. An illustrative instance occurred in September 2019 when several states froze assets of individual JNIM members, and the national government reportedly did not conduct raids on JNIM facilities or locations. Subsequently, 4 months later, in January 2020, presumed JNIM militants launched an attack on market goers in Solle, Burkina Faso, resulting in the death of ten civilians, injuries to others, abduction of women and children, and the ignition of two trucks.

7.5.1 TP-Rule TC-5

Attacks on civilians for reasons not linked to their identity occur in months in which:

- Up to 4 months earlier, foreign states or international institutions froze asset(s) of individual member(s) of the JNIM.
- Up to 4 months earlier, the government did not raid JNIM facilities and locations.

  Support = 0,33
  Probability = 98%, *Inverse Probability* = 64%, *Negative Probability* = 28%,
  Lift = 1.90

7.6 Targeting of Civilians When Foreign States or International Institutions Freeze Assets and the National Government Received Foreign Military Aid

The last condition coupled with the freezing of assets by foreign states or international institutions to predict the targeting of civilians is the occurrences when Mali, Burkina Faso or Niger receives foreign military aid. When both these conditions were met, we were able to predict the targeting of civilians sometime within the next 4 months.

7.6.1 TP-Rule TC-6

Attacks on civilians for reasons not linked to their identity occur in months in which:

- Up to 4 months earlier, foreign states or international institutions froze asset(s) of individual member(s) of the JNIM.
- Up to 4 months earlier, the government in the nation where the JNIM operates received foreign military aid.

   Support = 0,31
   Probability = 98%, *Inverse Probability* = 61%, *Negative Probability* = 29%, *Lift* = 1.90

7.7 Conclusions

The targeting of civilians by JNIM, irrespective of their beliefs, professions, or political backgrounds, has been a recurrent occurrence in Mali, Niger, and Burkina Faso over the past decade. The rationale behind this targeting is intricate and elusive, particularly in cases when citizens seem to be singled out for unspecified reasons. The capacity to forecast such targeting in advance would empower law enforcement to more effectively allocate resources, ensuring the protection of threatened communities and the preservation of people's livelihoods. In the following bullet points, we outline some conditions associated with the targeting of civilians, regardless of their identity. It is important to note that these relationships are correlational and not causative.

- When assets of (individual members of) JNIM are being frozen by foreign states or international institutions, targeting civilians tends to occur during the following four-month time interval.
- When the freezing of assets is combined with a second variable, targeting of civilians by JNIM tends to follow sometime during the next 4 months. These

"second" variables include: (top) leaders of JNIM not being under arrest or imprisoned, members of JNIM not being under arrest, no raids of the national government of JNIM locations or facilities, the national government not receiving military aid.
- JNIM appears to increase its targeting of civilians during the five-month window after foreign states impose a travel ban on Mali, Niger, or Burkina Faso.

While these correlations do not imply causation, they can contribute to the efforts against JNIM. These insights enable us to identify periods when civilian targeting is more likely, facilitating the strategic allocation of resources to counter various actions by JNIM. Redirecting manpower to guard against hit-and-run attacks or prevent abductions becomes possible. Furthermore, during periods of low likelihood of civilian targeting, resources can be directed toward assisting displaced individuals in rebuilding their lives and returning home. The task of defeating JNIM and rebuilding parts of the Sahel is formidable, but any tool that enhances our understanding of JNIM's actions holds significant value.

7.8 Predictive Model/Reports Results

Table 7.1 below shows the performance of our Northwestern Terror Early Warning System's (NTEWS) predictive model. We learned the predictive model using data up to the end of 2022 and then made monthly predictions during 2023. In other words, we predicted, at the beginning of January 2023 (using data till the end of December 2022), the attacks by JNIM sometime in the next month, the next 2 months, and so forth up to a 6-month lead time window. We then did the same thing at the beginning of February 2023 using the data till the end of January 2023. We used 6 classifiers on our dataset: SVM, KNN, Random Forest, Gaussian Naïve Bayes, Multinomial Naïve Bayes, and Logistic Regression, together with the late fusion module that combines the results of these predictors. The model predicts whether civilians will, or will not be targeted, within a given timeframe. For example, if the offset is 2 then the model predicts whether or not civilians will be targeted anytime during the next 2 months. Table 7.1 below depicts the results of our predictions compared to the ground truth observed after the predictions we made.

Table 7.1 Targeting of civilians results for our 2023 predictive reports

Targeting civilians						
Time period	1	2	3	4	5	6
Recall	82%	85%	91%	100%	100%	100%
Precision	88%	100%	95%	92%	96%	89%
AUC	84%	93%	90%	86%	92%	70%
F1	0.85	0.92	0.93	0.96	0.98	0.94

References

Al-Zawahiri (2013), General Guidelines for Jihad.
Anon (December 19, 2017 Tuesday). Al-Qaeda-linked Sahara group claims attack on Mali locals. *BBC Monitoring Africa—Political Supplied by BBC Worldwide Monitoring.*
Anon (January 12, 2019a Saturday). 12 civilians killed in jihadist attack in Burkina Faso. *Agence France Presse—English.*
Anon (November 7, 2019b). Dozens killed in attack on Canadian gold mine convoy in Burkina Faso. *China Daily—Africa Weekly.*
Anon (November 18, 2020 Wednesday). Timeline: Burkina Faso's growing jihadist threat. *Agence France Presse—English.*
Anon (June 20, 2022a Monday). At least 132 killed in jihadist militia attacks in northern Mali. *CE Noticias Financieras English.*
Anon (October 14, 2022b Friday). Ten killed in bus bombing in central Mali. *CE Noticias Financieras English.*
Clarke, M. (2019). Iraq: Terrorism and counter-terrorism in Iraq, 2003-2011. In A. Silke (Ed.), *Routledge handbook of terrorism and counterterrorism* (pp. 563–573). Routledge.
Jones, S. G. (2017). *Waging insurgent warfare.* Oxford University Press. https://doi.org/10.7249/CB543
Neumann, P. R., & Smith, M. L. R. (2005). Strategic terrorism: The framework and its fallacies. *Journal of Strategic Studies, 28*(4), 571–595. https://doi.org/10.1080/01402390500300923
Norman Cigar (2009). Al-Qa' ida's Doctrine for Insurgency, Dulles, Va.: Potomac Books Inc.
Richards, A. (2014). Conceptualizing terrorism. *Studies in Conflict & Terrorism, 37*(3), 213–236. https://doi.org/10.1080/1057610X.2014.872023

Chapter 8
Other Types of Attacks

NSAGs also have other ways of operating through which they try to achieve their objectives, which have not been discussed so far. This chapter therefore covers hit-and-run attacks, targeting of public transportation and sabotage of, for example, water purification plants. The TP-rules involving these variables shed additional light on how JNIM operates, what tactics they use, and how JNIM finances their operations. Robberies are often used as a way to obtain resources, such as vehicles or weapons. Robberies are often combined with the establishment of a (temporary) checkpoint on important roads. At these checkpoints, passers-by are often forced to hand over equipment and weapons. In addition, robberies take place in combination with larger-scale attacks on villages, often consisting of arson and destruction of properties.

Hit and run attacks are more common when, for example, an attack is made on a security installation or a military convoy. These attacks are often accompanied by IEDs or suicide bombings. Initially, JNIM conducted attacks mostly with explosive devices. Currently, armed assaults, assassinations, and ambushes are more common (Zimmerer, 2022).

Assassinations are a way for JNIM to control opponents, show their power towards the population or as an implementation of their ideology. Assassinations are therefore relatively common towards government officials, security professionals or civilians with a religious profession that does not correspond to the ideology of JNIMs. The following list shows some examples of these types of attacks.

- **1 July 2018:** JNIM carried out an attack in Gao, targeting a French and Malian patrol leaving four civilians dead and over a dozen others, including four French soldiers, wounded (Voanews, 2018).
- **3 April 2019:** Two peacekeepers were injured in a mortar and gun attack on a United Nations base in Kidal, in the desert north of Mali (The Defense Post, 2019)

- **5 July 2020**: Suspected JNIM militants attacked two communal transport vehicles in Kori Kori (Mali), killed four people, abducted one, and damaged the vehicles.
- **15 October 2020:** Presumed JNIM militants sabotaged electrical installations (solar panels and batteries) of a public primary school and removed a fence in the village of Delga, Burkina Faso (UNHCR, 2020)
- **16 March 2021:** JNIM militants sabotaged a telecommunications antenna near the village of Dounapen, Mali.
- **23 July 2022:** JNIM militants attacked a military base 15 kilometers outside of Bamako using two car bombs leading to six people wounded (Al Jazeera, 2022).
- **24 October 2022:** An attack on the military barracks in Djibo, Burkina Faso left at least 10 people dead (Crisis24, 2022).

Figure 8.1 depicts the frequency of hit-and-run attacks, assassinations, and robberies per month over the years, indicating a steady increase since the official creation of JNIM in 2017. Figure 8.2 showcases the total casualties resulting from hit-and-run attacks and assassinations per month throughout the analyzed period, along with the frequency of robberies per month during the same timeframe.

The array of attacks orchestrated by JNIM has played a destructive role in the group's terror campaign. Each successful hit-and-run operation contributes to the erosion of morale among both civilians and soldiers. This chapter specifically delves into TP-Rules designed to predict the occurrence of these attacks. Additionally, it explores the sabotage tactics employed by JNIM, allowing the group to influence the availability of basic necessities. The final rule outlined in this chapter focuses on

Fig. 8.1 No. of months per year with assassinations, hit and run attacks and robberies

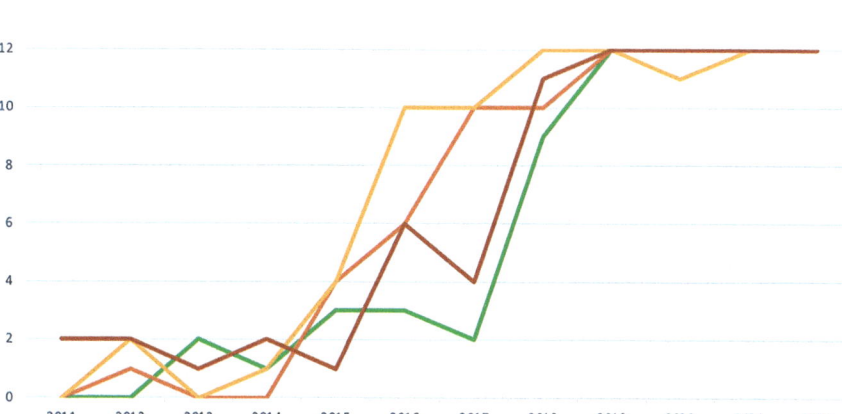

Fig. 8.2 Number of instances of assassinations, hit and run attacks, and robberies per month

the targeting of public transportation facilities, predominantly buses. The subsequent section will delve into the variables integral to these TP-Rules.

- *The national government received international military aid.* During months when the national government of the state where JNIM operates (Mali, Niger or Burkina Faso) received international military aid, we were able to devise TP-rule for predicting hit and run attacks sometime during the next 6 months.
- *National security forces did not use sexual violence.* In months when national security forces did not use any sexual violence against civilians, and the national government received international military aid, we were able to derive TP-rules predicting hit and run attacks sometime during the next 6 months .
- *The leadership of JNIM is not split or fractured.* We were able to derive TP-rules predicting hit and run attacks during the 5 month window immediately after months when the leadership of JNIM was not split or fractured, and the national government was receiving international military aid.
- *A travel ban was placed on Mali, Niger, or Burkina Faso.* By examining months when a travel ban was placed by foreign countries on Mali, Niger or Burkina Faso combined with the freezing of assets of JNIM, we were able to derive TP-rules to predict sabotage sometime during the next 6 months.
- *JNIM addresses the general public in its communications.* By examining months when JNIM addresses the general public in its communications and months when a travel ban was placed on a country where JNIM operates, we were able to predict targeting of public transportation sometime during the succeeding 6 months.

8.1 Hit and Run Attacks when the National Government Receives International Military Aid

This chapter will focus on a few rules that can be used to predict several different attacks and events. The first of these rules is TP-Rule MS-1 (Miscellaneous), which can be used to predict when JNIM will carry out hit and run attacks. For this rule to be viable, the national governments of Mali, Burkina Faso or Niger must have received international military aid. This aid can take different forms, from international security forces present in Mali, Burkina Faso or Niger to receiving weapons. The former being the case from April 2013 when the United Nations Multidimensional Integrated Stabilization Mission in Mali started. Rule MS-1 can be viewed below.

8.1.1 TP-Rule MS-1

Hit and run attacks occur in months in which:

- Up to 6 months earlier, the national government received international military aid.

    Support = 0,49
    Probability = 64%, *Inverse Probability* = 94%, *Negative Probability* = 12%, *Lift* = 1.25

8.2 Hit and Run Attacks when the National Government Receives International Military Aid and Government Security Forces Did Not Use Sexual Violence

Similar to TP-rule MS-1, Rule MS-2 also investigates hit-and-run attacks, utilizing instances where the national government receives international military aid as a variable. Additionally, this rule incorporates the absence of sexual violence by (national) security forces as another variable to predict hit-and-run attacks 6 months in advance. Both conditions aligned in November 2019 when security forces reportedly refrained from using sexual violence, and Mali was receiving foreign military aid. Consequently, 6 months later, in May 2020, presumed Katiba Macina (JNIM) militants attacked the Dogon village of Tile, resulting in the reported death of seven individuals, including a Dan Na Ambassagou[1] chief, a pregnant woman, and five children, with the village set ablaze. The details of TP-rule MS-2 are presented below.

---

[1] Dan Na Ambassagou is an ethnic militia in Mali

8.2.1 TP-Rule MS-2

Hit and run attacks occur in months in which:

- Up to 6 months earlier, the national government received international military aid.
- Up to 6 months earlier, national security forces did not use sexual violence.

    Support = 0,47
    Probability = 66%, *Inverse Probability* = 92%, *Negative Probability* = 15%, *Lift* = 1.29

8.3 Hit and Run Attacks when the National Government Receives International Military Aid and the Leadership of JNIM Was Not Split or Fractured

TP-rule MS-3 is the final rule within this segment of this chapter, specifically designed to forecast hit-and-run attacks orchestrated by JNIM. Similar to the previous rules, this one also incorporates the variable of the national government receiving foreign military aid. However, it introduces a different second variable: the status of the leadership within JNIM. When the leadership of JNIM remains unified, without any splits or fractures, and concurrently the national government of the state where JNIM operates receives international military aid, we have successfully predicted hit-and-run attacks occurring sometime during the subsequent five-month time window.

8.3.1 TP-Rule MS-3

Hit and run attacks occur in months in which:

- Up to 5 months earlier, the national government received international military aid.
- Up to 5 months earlier, the leadership of JNIM was not split or fractured.

    Support = 0,48
    Probability = 65%, *Inverse Probability* = 94%, *Negative Probability* = 11%, *Lift* = 1.7

8.4 Sabotage When a Travel Ban Was Placed on the State Where JNIM Operates and there Was no State of Emergency

This segment of the chapter explores Rule MS-4, designed to predict sabotage orchestrated by JNIM. To satisfy Rule MS-4, a travel ban must be imposed on the state where JNIM operates. The second condition is met when there is no state of emergency declared by the national government of the respective states. For instance, in March 2021, presumed Katiba Macina (JNIM) militants sabotaged a telecommunications antenna in the village of Dounapen, Mali. This incident occurred 6 months after a travel ban was imposed on Mali by several foreign states in September 2020, and no state of emergency was declared.

8.4.1 TP-Rule MS-4

Sabotage occurs in months in which:

- Up to 6 months earlier, a travel ban was placed on the country where JNIM operates.
- Up to 6 months earlier, there was no state of emergency in the countries where JNIM operates.

   Support = 0,10
   Probability = 66%, *Inverse Probability* = 70%, *Negative Probability* = 5%, *Lift* = 4.6

8.5 Targeting of Public Transportation Facilities When JNIM Addresses the General Public and a Travel Ban Was Placed on the State Where JNIM Operates

The final rule explored in this chapter centers on the targeting of transportation facilities, with JNIM frequently directing attacks toward public transportation, particularly buses, in their anti-terrorism efforts. However, not all of these attempts are successful; some are prevented, while others may lack the efficacy to cause injuries or fatalities. TP-Rule MS-5 offers predictions for anticipating instances of public transportation targeting. This rule enables predictions to be made during the subsequent six months' time frame, particularly if JNIM is reported to be addressing the general public in their communications or if a travel ban is imposed on Mali, Niger, or Burkina Faso during a given month. The details of Rule MS-5 can be found below.

8.5.1 TP-Rule MS-5

Targeting of public transportation facilities occurs in months in which:

- Up to 6 months earlier, JNIM addressed the public.
- Up to 6 months earlier, a travel ban was placed on the country where JNIM operates.

  Support = 0,07
  Probability = 53%, *Inverse Probability* = 66%, *Negative Probability* = 4%, *Lift* = 4.84

8.6 Conclusions

The diverse array of attacks perpetrated by JNIM significantly contributes to the group's reign of terror. The deliberate targeting of public transportation facilities enables the group to disrupt communities and restrict the freedom of movement for individuals. Similarly, the numerous hit-and-run attacks carried out by the group in various villages or military installations, often accompanied by arson and looting, further increase the atmosphere of fear. JNIM's periodic targeting of telecommunication masts adds to the disruption, leading to reduced interconnectivity among individuals. Outlined below are descriptions of some events linked to these various terror attacks. It is crucial to emphasize that these relationships are correlative and not causative, providing insights into the patterns of JNIM's actions without implying direct causal connections.

- In months when the national government of the state where JNIM operates receives foreign military aid, hit-and-run attacks appear to occur during the subsequent six-month period.
- Additionally, when the national government receives military aid and security forces refrain from using sexual violence against civilians, hit-and-run attacks seem to occur during the following six-month time interval.
- Moreover, if the national government receives military aid and the leadership of JNIM is split or fractured, hit-and-run attacks appear to occur within the next 5 months.
- JNIM appears to engage in sabotage when, up to 6 months earlier, a travel ban was imposed on the state where they operate, and there was no state of emergency.
- Furthermore, JNIM seems to target public transportation facilities in months when, during the preceding 6 months, a travel ban is placed on the country where they operate, and they address the general public in their communications.

While the connections outlined above do not imply causation, they do offer insights into the circumstances surrounding some of JNIM's attacks. Understanding these circumstances can enhance the allocation of resources and assets to combat them effectively. For instance, anticipating hit-and-run attacks allows for the implementation of security alerts and the deployment of additional personnel to better safeguard military installations and villages. The challenge of fighting an asymmetrical conflict has proven formidable in the Sahel, echoing experiences from the twentieth and twenty-first centuries. The hope is that the rules and relationships uncovered can provide some assistance in making this challenging task slightly more manageable.

8.7 Predictive Model/Reports Results

Tables 8.1, 8.2 and 8.3 below show the performance of our Northwestern Terror Early Warning System's (NTEWS) predictive model. We learned the predictive model using data up to the end of 2022 and then made monthly predictions during 2023. In other words, we predicted, at the beginning of January 2023 (using data till the end of December 2022), the attacks by JNIM sometime in the next month, the next 2 months, and so forth up to a 6-month lead time window. We then did the same thing at the beginning of February 2023 using the data till the end of January 2023. We used 6 classifiers on our dataset: SVM, KNN, Random Forest, Gaussian Naïve Bayes, Multinomial Naïve Bayes, and Logistic Regression, together with the late fusion module that combines the results of these predictors. The model predicts whether hit and run attacks on security forces (Table 8.1), sabotages (Table 8.2), and targeting of public transport (Table 8.3) will, or will not be conducted, within a given timeframe. For example, if the offset is 2 and the attack under analysis is sabotage, then the model predicts whether or not sabotage will occur anytime during the next 2 months. Tables 8.1, 8.2 and 8.3 below depict the results of our predictions compared to the ground truth observed after the predictions we made.

Table 8.1 Hit and run attacks results for our 2023 predictive reports

Hit & Run attacks						
Time period	1	2	3	4	5	6
Recall	94%	89%	95%	95%	90%	95%
Precision	89%	94%	100%	100%	100%	100%
AUC	90%	91%	97%	97%	95%	97%
F1	0.91	0.91	0.97	0.97	0.95	0.97

Table 8.2 Sabotage results for our 2023 predictive reports

Sabotage						
Time period	1	2	3	4	5	6
Recall	100%	86%	78%	78%	90%	80%
Precision	100%	100%	88%	100%	81%	100%
AUC	100%	93%	87%	89%	90%	90%
F1	1	0.92	0.82	0.88	0.86	0.89

Table 8.3 Targeting of public transportation results for our 2023 predictive reports

Targeting public transportation						
Time period	1	2	3	4	5	6
Recall	60%	86%	89%	80%	91%	100%
Precision	60%	50%	67%	73%	91%	86%
AUC	76%	80%	85%	83%	93%	94%
F1	0.6	0.63	0.76	0.76	0.91	0.92

References

Al-Qaeda affiliate claims deadly Mali attack. (2022). Al Jazeera. Retrieved April 18, 2024, from https://www.aljazeera.com/news/2022/7/23/al-qaeda-affiliate-group-claims-deadly-mali-attack

Burkina Faso: Armed attack leaves at least 10 dead and 50 others injured in Djibo, Sahel Region, Oct. 24. (2022). Burkina Faso: Armed Attack Leaves at Least 10 Dead and 50 Others Injured in Djibo, Sahel Region, Oct. 24 | Crisis24. Retrieved April 18, 2024, from https://crisis24.garda.com/alerts/2022/10/burkina-faso-armed-attack-leaves-at-least-10-dead-and-50-others-injured-in-djibo-sahel-region-oct-24

Mali: UN peacekeeper injured in attack on Kidal MINUSMA base. (2019). Retrieved April 18, 2024, from https://www.thedefensepost.com/2019/04/04/mali-attack-kidal-un-minusma/

News, V. O. A. (2018, July 2). Al-Qaida Affiliate Claims Responsibility for Mali Attack. Voice of America. https://www.voanews.com/a/al-qaida-affiliate-claims-responsibility-mali-attack/4463606.html

UNHCR (2020) Evaluation of UNHCR'S response to multiple emergencies in the central sahel region: Burkina Faso, Niger, Mali https://www.unhcr.org/sites/default/files/legacy-pdf/63c95f1c4.pdf

Zimmerer, M. (2022). Terror in West Africa: A threat assessment of the new Al Qaeda affiliate in Mali. Critical Studies on Terrorism, 12(3), 491–511.

Chapter 9
Reflections & Implications for Military Decision Making

It is an old military wisdom that to defeat adversaries, one must understand and defeat their strategy rather than their armed forces. The underlying argument is that victory does not merely lie in defeating sheer force, but in outmaneuvering the opponent's intentions (Tzu, 2000). This holds equally true when facing a state or non-state actor. By comprehending their ends, means and ways, one can strategically counter the opponent's moves while exploiting perceived vulnerabilities. Defeating an adversary thus extends beyond a physical confrontation: it necessitates undermining his political-strategic foundations, disrupting his alliances, and rendering the adversary's methods ineffective. This makes recognizing and understanding the adversary's strategy paramount for victory.

By using JNIM as a case study, this book has offered a model-based approach for analyzing and comprehending the activities of terrorist and insurgent organizations. Activities are the 'ways' in a strategy and understanding how these contribute to the 'ends' thus forms a crucial part in understanding the adversary's strategy. The primary objective of the analytic method in this book is to furnish timely and precise decision support for policy makers and intelligence professionals dealing with these adversaries. The emphasis of this chapter lies in assessing the model's capacity, as presented in this book, to facilitate the decision-making process, particularly in a military context. In the following paragraphs, we employ a model-based approach to the intelligence process to gauge the model's timeliness and accuracy, while also endeavoring to incorporate case-specific enhancements.

9.1 The Model-Based Intelligence Cycle

In the realm of intelligence and defense communities, it is standard practice to utilize a structured analytical framework commonly referred to as 'the intelligence cycle' to comprehensively grasp the intentions and capabilities of adversaries. In recent years, several adaptations of this cycle have emerged, explicitly integrating data science and other quantitative models. In this context, we apply Waltz's model-based intelligence process to evaluate and assess the utility of the data-centered scientific model of JNIM presented in this book within military organizations (Waltz, 2014). Waltz's process consists of seven steps, which we'll use as a framework to discuss the utility and possible extensions of the model of JNIM presented in this book:

- Step 1: Define the intelligence problem.
- Step 2; Identify information needs and gaps, subsequently define the required knowledge (and the purpose of the to be created models).
- Step 3: Plan, task and collect data about the situation at hand, using quantitative data modeling and qualitative data analysis.
- Step 4: Structure the analytic process, synthesize hypotheses and marshal data to hypotheses.
- Step 5: Perform target modeling by decomposing the situation into abstracted components (elements and systems), identify theories of causality (the basis of dynamic models) and select and specify model representations, construct and validate the model for the right purpose.
- Step 6: Infer, synthesize and evaluate alternative hypotheses and explore the structure and dynamics of systems to test hypotheses.
- Step 7: Apply analytic judgement, articulate the assessments, supporting rationale and confidence level based on the scope and quality of supporting information.

9.1.1 Defining the Intelligence Problem (Phase 1)

First, given an unfolding security situation, such as a non-state armed group conducting attacks and subverting a government, there is a need to 'define the intelligence problem'. In the case of JNIM in the Sahel region, one can imagine that policy and decision makers need intelligence on a variety of factors, such as the structure and leadership of the organization (if any), the financing and recruitment systems in place, but most importantly its intentions, capabilities and the conditions and the likelihoods under which the group will undertake activities to attain its goals. In short, the most critical requirement is to understand the adversary's strategy by observing its actions. Understanding the strategy exposes possible vulnerabilities or venues of approach to counter the adversary, in this case JNIM.

9.1.2 Identifying Intelligence Needs Gaps and Defining Required Knowledge (Phase 2)

Second, critical gaps have to be identified in both the information on the respective organization and the knowledge and understanding of the organization. According to Waltz (2014), this includes, amongst other factors, the identification of the key variables at stake, ultimately defined in our custom-made codebook. This phase also includes defining the boundary of the social target system, i.e., which individuals and groups to include or exclude in the analysis, and what open and closed source datasets to collect to identify the key variables. The second phase also entails identifying ways to increase knowledge about the phenomenon at hand. What computational models, such as agent-based or social network analytic models, are already available and deployed? If in other situations or contexts TP-models, such as the one presented in this book, have been used to analyze certain organizations, what were the outcomes, benefits and shortcomings? How can they be utilized in this context, and improved?

In our model of JNIM, the variables, as defined and described in the codebook, were pre-defined by the developers of the model in a political rather than military context. Since the variables used in the current codebook are defined from a political perspective, they also offer insight into more politically driven information or knowledge gaps. This means that the information gaps which could concern a military decision maker, such as insight into the use of certain weapons by an insurgency, are currently not adequately addressed. Military analysts could enhance the codebook used in this book by providing input on variables to include or exclude based upon their ability to identify military information or knowledge gaps and their expertise and knowledge of military operations. The exclusion of relevant non-military variables can impede the model's utility for understanding an adversary's strategy. Other non-military variables that encompass political and organizational dynamics, financial-economic interests, cultural and ethnic factors, and collective and personal historical experiences and grievances often shape an adversary's decision-making process and strategic objectives. By excluding such variables, the model currently provides a more appropriate response to political or strategical defined intelligence gaps rather than intelligence needs that may be posed by a military decision maker. Identifying information needs thus also entails a detailed discussion between subject matter experts of the unfolding situation at hand and model developers on the other, to identify the specific intelligence gaps and the subsequent variables needed. It is of paramount importance to develop a codebook using variables which are identified by military subject matter experts (SMEs) and model developers in concert, preventing inaccurate assessments of the adversary's intent, capabilities, and likely courses of action and the associated vulnerabilities. And above all, to enable the model to ultimately provide answers that fulfill the information needs of a military decision maker.

9.1.3 Collection (Phase 3)

During the collection phase, the information needs that are defined in earlier stages are translated into a collection plan. In general, the collection plan tasks the various intelligence collection sources (HUMINT, SIGINT, etc.). The quantitative data is processed, stored and tagged. The event (i.e. attack) data for the behavioral model in this book is gathered purely from open-source data such as ACLED database, while LexisNexis and BBC monitoring were used to gather the independent (i.e. environmental) variables. Enriching open-source data with closed source data collected by intelligence organizations could be valuable. Intelligence sources provide information through human intelligence, signals intelligence, imagery intelligence, etc., giving insights and perspectives on adversaries, operations, and intentions that are hard to come by using open-source data alone. Information provided by intelligence, for instance, enables counterinsurgents to uncover the intention behind activities, and discern strategy driven patterns or internal organizational dynamics that may not be discernible from open sources alone. Moreover, closed source data can potentially corroborate or refute open-source data, enhancing the accuracy and reliability of assessments. By integrating intelligence-derived insights with open-source information, analysts can potentially gain a more nuanced understanding of an adversary, enabling better informed decision-making and proactive responses to evolving threats.

Consider the following example to illustrate this point. In November of 2023, Kidal was captured by the FAMA (Malian Armed Forces) and Russian PMC Wagner. This city was a key stronghold for the Coordination of Movements of Azawad (CMA) and represents the political-administrative and cultural heartland of the Tuareg-led rebellion. The CMA suffered a significant blow with the loss of Kidal, as it undermined their territorial control and weakened their bargaining position in negotiations with the Malian government. Additionally, the renewed government presence in Kidal could disrupt the CMA's influence over local governance and resource management in the region. Although the (open-source analysis of the) capture of Kidal might lead to a conclusion of a strengthening of the position of the government and a weakening of that of the insurgents, another conclusion could be that the capture of Kidal may inadvertently have politically strengthened JNIM. With the CMA weakened, JNIM could exploit the power vacuum and expand its influence, presenting itself as the sole organization able to stand up for the interests of the Tuareg people in Mali and the wider Sahara-Sahel region. Whether this is actually the case is difficult or even impossible to find out through open sources alone. It is likely that these possible dynamics can only be identified by obtaining intelligence from within the organization or by monitoring indicators that are difficult to identify. The interpretation of the rules presented in this book should therefore entail a collaborative effort between subject matter experts (SMEs) and data analysts to derive suitable insights from the model's outcomes. Such a collaboration could help ensure that the complexity behind the behavior of JNIM is properly addressed.

Additionally, this type of collaboration can either make the variables used in the model more comprehensive or indicate the need to create new variables.

9.1.4 Synthesis (Phase 4)

During this phase, subject matter experts can meet in workshops with model builders to develop hypotheses and models about the situation and the organization at hand. It is during this session that the shortcomings mentioned in the previous subsections, i.e. that of appropriate variable selection and corresponding data collection, can be fleshed out. During this phase, the codebook as used in this book and developed by civilian computer scientists and political scientists should be reviewed and adapted to the military operation(s). Subject matter experts possess specialized knowledge, experience, and understanding of the domain under study and the organization at hand, enabling them to contextualize findings within relevant frameworks, theories, own experiences and real-world scenarios. During this phase, subject matter experts can apply their domain-specific expertise to interpret the implications of quantitative findings and help formulate the appropriate parameter settings for the models used. While the codebook used in this book was developed in cooperation with predominantly experts with a background in political science, it does provide certain variables which could be useful in a military context. For example, some rules regarding the targeting of security professionals and/or installations describe how the absence or decrease in executions by security forces positively correlates with the absence of targeting of security forces (see Sect. 6.4, TP-rule 6–5).

However, taking on a military perspective could lead to the addition of new variables using other datasets. An example is the possible correlation between the activities of improvised explosive devices (IEDs) and weather conditions. In the period of 2015–2016, restricted operational data and firsthand experiences of one of the authors indicated a notable correlation between the onset of rain and an intensified activity of planting IEDs along semi-paved roads (Provoost, Deputy Commander Special Operations Land Task Group MINUSMA, 2015–2016). The rainy season softened the ground, allowing insurgents to relatively easily bury IEDs beneath the surface without arming them. As the soil hardened again after the rain season had ended, the prepared devices became exceptionally challenging to detect, especially because local traffic could continue using these roads. As a result, IEDs could be activated weeks or months after they had been put in place, when they were far more difficult to detect than if they were actually emplaced just prior to their usage. This *modus operandi* implies a correlation between the rainy season and a temporal heightening in IED attacks. The example indicates a need for variables that include weather effects to be included in the model, if analysts wish to have a better understanding of JNIMs tactics.

Another example concerns identifying the ethnicity of victims of violence. This is essential for discerning whether fueling sectarian violence is part of an insurgent group's strategy. Ethnic differences can be used to create and exacerbate fault lines in a population, with insurgents either exploiting existing tensions or creating tensions as a source of conflict to further their own agendas. By understanding the ethnic composition of victim populations, analysts and policymakers can trace patterns of targeted attacks and discern whether they align with broader sectarian objectives. Recognizing these patterns can help shed light on insurgents' motives, strategies, and alliances, enabling more effective counterinsurgency efforts. Moreover, understanding the ethnic dimensions of violence is crucial for conflict resolution initiatives (Kaufmann, 1996). Without a comprehensive understanding of ethnicity's role in violence, efforts to mitigate sectarian tensions and build sustainable peace are likely to fall short. Therefore, identifying the ethnicity of victims is an imperative for better understanding the adversary's strategy in this type of conflict.

9.1.5 Target Modeling (Phase 5)

During this phase, the analytic methodologist, in our case the developer(s) of the TP-model and the codebook, collaborate with the intelligence analyst(s) to make a selection of relevant variables gathered during the synthesis phase. The selection of a model suitable for the intelligence problem at hand is crucial in this phase. In our case, the model should be general, but also applicable to insurgencies. That is, it should represent a widely supported theory of insurgency operations. In this book, the target model consists of a behavioral model of JNIM in the form of temporal-probabilistic rules. The rules consist of independent variables, pre-determined by computational and political analysts. While the independent variables are diverse, consisting of several social, cultural, political, economic and military variables, they are not directly reconstructable from a widely supported theory of insurgencies. To further improve the utility of the current model for predicting behavior of JNIM, variables could be defined using historical data on Islamist insurgencies, perhaps by into account, Mao's phases as a guiding theory. Furthermore, by analyzing patterns and trends from past insurgencies, the model can identify common trajectories and benchmarks associated with progressing through Mao's phases and the resulting changes in violence types. Factors such as the socio-political landscape, levels of external support, government responses, and insurgent network resilience play crucial roles in shaping the duration of each phase. Comparing historical examples of Islamist insurgencies allows the model to discern similarities and differences in these factors, enabling informed estimations of phase durations. While each insurgency is unique, historical precedents provide valuable reference points for understanding conflict evolution dynamics. The model's ability to draw parallels and extrapolate trends from past experiences enhances its capacity to forecast the approximate duration of each phase in Mao's model, aiding in counterinsurgency efforts by prompting renewed strategic planning, potentially requiring changes in security personnel deployment, resource allocation, and political efforts.

9.1 The Model-Based Intelligence Cycle

9.1.6 Inferential and Exploratory Analysis (Phase 6)

During the sixth phase the intelligence team engages in both inferential and exploratory analyses to assess the hypotheses generated in phase four. To test these hypotheses, the analytic methodologist creates various versions of computational models. These models help compare empirical data with what we would expect if each hypothesis were true. The intelligence team also simulates different scenarios to see which actions could prove or disprove each hypothesis. As more evidence is presented, the team is able to compare it with expectations presented in their models.

The model presented in this book could play a similar role in phase 6 as described by Waltz. The rules as presented in Chapters 4 through 8 provide a basis on which hypotheses regarding the behavior of JNIM can be refined, confirmed or disproved. At the same time, the rules may point to entirely new hypotheses. In this case, for example, it is conceivable that rules could point to the transition from the phases of Mao, as discussed earlier in this book. Hypotheses about the moment of transition or the specific layout of a subsequent phase can be defined in response to rules generated by the model. An example are the rules defined regarding attacking public sites in Chap. 5. In a number of these rules, variables are used that indicate the failure of JNIM to communicate (strategically) about actions or claim responsibility for these actions. This is positively correlated with an increase in attacks on public sites. At the same time, an increase in attacks on public sites correlates with phase two/three as described in Sect. 2.2.2, as the attacks do not focus on "clean targets" but rather on smaller cities or public sites.

To fully meet the description of phase 6 of Waltz (2014), a second model should be used that supports a secondary hypothesis. This second model could potentially be a variation of the current model in which other variables are defined. For example, the hypotheses could be defined using Mao's phases, with the two different models deriving their variables from different phases. Therefore, the completion of stage 6 is also an interplay between the SME's and analysts, where both can reinforce each other in the analysis process by discussing hypotheses and, consequently, defining variables.

9.1.7 Judgment and Reporting (Phase 7)

One of the core elements of intelligence analysis is the ability to communicate the results and subsequent conclusions in an understandable way to enable decision makers to be able to make the right decisions. This is described by Waltz (2014) in phase seven, the last phase of his description of the intelligence cycle. Waltz emphasizes the need to draw understandable and insightful conclusions from the model, without distracting from the complexity of both the NSAG studied as well as the model itself. Waltz (2014) also highlights the importance of expressing confidence in the model used by the analyst briefing the results. This confidence is based upon the data used in the model, the theories of structures and causality applied, and the internal or external validity of the model. In the case of the model described in this book, it is essential to note that the TP-rules present correlations, not causal relationships. Assigning real-world behaviors of a NSAG to these rules can, by confirming external validity, render the model outcomes actionable for a decision-maker. The knowledge and experiences of a subject matter expert are once again crucial in this context. The outcomes of the model can support hypothesizing enemy courses of action. These adversarial courses of action are taken into account in our decision-making. TP-rules, as found by the model, can serve as indicators for the most likely or most dangerous courses of action. For example, JNIM's behavior of claiming attacks could indicate future attacks on security forces, information that can be crucial for a military decision-maker (see TP-rule P6–3 and P6–4 in Chap. 6) to be aware of. Ultimately, it is imperative for the intelligence team to ensure that this information, whether derived from the model or SME analysis, is presented to the decision-maker with the highest possible certainty and in a timely manner. The model presented in this book can add value to this process by presenting alternative perspectives or supporting (new) enemy courses of action.

Furthermore, a machine learning model generating behavioral rules of adversarial groups running in the background of an ongoing military operation serves as a dynamic and continuous knowledge base, offering valuable insights in trends, predicting possible future developments and providing guidance to newly arrived military personnel lacking prior in-theater experience. By leveraging real-time data, historical records, and contextual information, the model can rapidly analyze the evolving operational environment, identify patterns, and generate actionable recommendations tailored to specific mission objectives and situational dynamics. Such a machine learning model potentially enhances situational awareness, and operational effectiveness of military forces, fostering a culture of learning, innovation, and continuous improvement within the operational theater. Furthermore, it mitigates the loss of knowledge occurring when military units relieve each other in place, ensuring that critical insights and lessons learned are better transferred across different troop rotations. Upon detecting indicators (or TP-rules in the context of the model presented in this book) of impending challenges or opportunities, the model can prompt military staff to initiate planning processes aimed at mitigating risks, exploiting opportunities, and optimizing operational readiness. This proactive approach allows military units to stay ahead of events, ensuring that plans are developed and resources allocated in advance, rather than in reaction to when they unfold. By leveraging the predictive capabilities of machine learning, military organizations can enhance their agility, resilience, and responsiveness to dynamic and uncertain operational environments, ultimately improving their ability to achieve mission success.

Phase 7 is the final phase described by Waltz (2014), but this does not mean that the intelligence process ends here. As previously stated, the process occurs in the form of a continuous iterative cycle. This means that it is necessary to continue testing the model using real-world scenarios and refining or adding variables as needed.

The model thus adds value throughout the entire cycle and in the end product of this cycle: supporting (military) decision-makers in making the right decisions.

9.2 The Value of Adding Geo-Spatial Data

The current model's ability to process and analyze data over time is valuable, but its efficacy would be significantly enhanced by incorporating associated geographic locations into its framework. Military thinking about events is inherently characterized by coupling activities with both time and space, as geographic location plays a crucial role in operational planning, decision-making, and situational awareness. By integrating geographic data into the model, military planners can gain a more comprehensive understanding of the events in the operational environment and anticipate how events may unfold spatially over time. This spatial-temporal perspective enables military units in advance to identify key chokepoints, plan maneuver routes, and allocate resources more effectively, maximizing operational effectiveness and minimizing vulnerabilities. By combining data processing with spatial analysis capabilities, the model can provide military decision-makers with actionable insights that are not only timely but also spatially relevant, enhancing their ability to counter threats and attacks.

Furthermore, in insurgency and counterinsurgency thinking, the interpretation of violence in a specific area serves as a critical indicator of the ongoing power dynamics and control exerted by belligerent factions. David Kilcullen, a renowned counterinsurgency expert, discusses the relationship between violence and control in insurgency environments (Kilcullen, 2010, p.57). In his writings, he frequently highlights the significance of interpreting violence as an indicator of contested or controlled areas within an insurgency context. When violence is prevalent in a particular area, it is often interpreted as a sign of contestation among competing groups seeking to establish dominance or influence over the territory. The escalation of violence signifies the struggle for control, with insurgents, government forces, and other armed actors vying for territorial supremacy and popular support. Conversely, the absence or limited presence of violence is commonly interpreted as the area being under the control of one of the belligerents, whether it be the government, insurgent groups, or local militias. This relative calm suggests a degree of stability and authority established by the controlling faction, allowing for governance, social control, and the implementation of strategic objectives. Understanding the nuanced relationship between violence and control in insurgency contexts is essential for assessing the fluidity of power dynamics, predicting potential flashpoints, and devising effective counterinsurgency strategies aimed at stabilizing contested areas and consolidating control.

Taking Kilcullen's hypothesis as a basis, the inclusion of geospatial data alongside event information within an insurgency context serves as a powerful tool to visualize and discern areas that are contested or under the control of various factions. By overlaying event data, such as instances of specific types of violence against specific types of targets, onto geographic maps, analysts and decision-makers gain a spatial understanding of the dynamics of the ongoing conflict and can identify clusters of violence, assess the distribution of incidents, and discern trends in specific areas. This spatial visualization aids in highlighting contested zones, where different belligerents are vying for control, influence, or resources. Clusters of violence may indicate what the adversary perceives as strategic locations, key supply routes, or areas of ethnic or political significance. Conversely, areas with limited or no reported events may indicate zones under the effective control of one faction, whether government forces, insurgent groups, or local militias. Geospatial visualization, derived from the fusion of temporal and spatial data, not only highlights patterns and trends in conflict dynamics but also aids in identifying critical nodes, vulnerabilities, and opportunities, enabling counterinsurgents to anticipate hotspots, formulate targeted measures and allocate resources effectively to stabilize contested regions and reinforce control in contested areas or those already under control. As such, the integration of geospatial data enhances situational awareness, facilitates informed decision-making, and enables more effective allocation of resources.

9.3 Defining Variables and Mao's Framework

Defeating JNIM requires understanding and countering its strategy effectively, assuming it follows Mao's three-phase model. However, accomplishing this necessitates more detailed and differentiated variables in the model to accurately assess the phase of the insurgency. By focusing on high-visibility targets and maximizing casualties in phase 1, insurgents aim to establish their presence, undermine the legitimacy of the state, and lay the groundwork for further destabilization efforts as the insurgency progresses. In phase 2 of an insurgency, civilian casualties may occur as a result of collateral damage resulting from an increasing number of guerrilla attacks or an intentionally provoked overreaction by the counterinsurgent forces. As the conflict intensifies and insurgents become more entrenched within civilian areas, counterinsurgency operations often necessitate the use of force in densely populated areas where insurgents operate. In such environments, distinguishing between combatants and non-combatants becomes increasingly challenging, leading to inadvertent civilian casualties as collateral damage. Moreover, in response to insurgent attacks or perceived threats, counterinsurgency forces may overreact or employ heavy-handed tactics, resulting in civilian casualties due to indiscriminate use of force or disproportionate responses. These casualties, whether caused by collateral damage or an overzealous counterinsurgency response, are often intended by the insurgents as they can erode civilian trust in the government and security forces, fueling resentment and providing recruitment propaganda for the insurgents.

9.3.1 From Mao to Variables

These are examples of the complexity in behavior exhibited by JNIM. At the same time, these examples also illustrate how difficult it can be to define this complexity in variables. Civilian casualties, for example, can result from targeted actions or collateral damage. However, the current codebook used in this research does not distinguish between the two. Rules regarding civilian casualties may therefore be biased or require further clarification. The same applies to the targeting of civilians. For example, there is no separate variable in this book for targeting civilians based on a specific ethnic background, despite this being the case in practice. That said, our prior book on applying this framework to the case of the terror group Lashkar-e-Taiba does consider targeting of civilians on the basis of ethnicity as a dependent variable (Subrahmanian et al., 2013). Currently, such cases are coded in this book under a kind of 'collective' variable, where all instances of targeting civilians are coded, making it difficult to determine the specific reasons for each case. By integrating additional variables and refining definitions of variables, analysts can better determine where JNIM stands within Mao's framework. With more nuance and differentiation in the outcome variables, the model can provide valuable insights from quantitative data analysis on how JNIMs strategy might evolve in the near future. By anticipating potential shifts in tactics and objectives, security forces can adapt their approach accordingly, disrupting JNIM's efforts and working towards neutralizing the insurgency more effectively.

Distinct counter-strategies are suited to counter the evolving tactics, use of violence, and intermediate objectives associated with each phase of Mao's model of insurgency effectively. During phase 1, characterized by clandestine organization and preparation, counterinsurgency efforts should prioritize intelligence gathering, surveillance, and disruption of insurgent networks (Galula & Nagl, 1964). By dismantling clandestine cells and disrupting recruitment efforts, counterinsurgents can prevent the insurgency from gaining momentum and overcoming the dreaded collective action problem (Jones, 2016). In phase 2, marked by guerrilla warfare and an increase in violence and attacks, counterinsurgency strategies should focus on gaining and retaining control over a population through a comprehensive approach of targeted development projects, provision of security, and promotion and strengthening of suitable governance structures that address grievances, diminish socio-economic inequality and mitigate incentives to support the insurgents (Galula & Nagl, 1964). Finally, in phase 3, when the insurgency transitions to conventional warfare, counterinsurgency efforts must prioritize preventing a general collapse of state structures, engage in decisive military action to degrade conventional insurgent capabilities, retain and regain territorial control, and isolate fielded insurgent forces from their support networks (Galula & Nagl, 1964). By tailoring counter-strategies to the specific dynamics of each phase, counterinsurgency tactics can effectively undermine the insurgent organization's intermediate objectives, disrupt its operations, and ultimately defeat its strategy.

9.3.2 Including Adaptability and Flexibility

However, it should be noted that insurgencies are often characterized by adaptability and fluidity (Farrell, 2018; Kilcullen, 2010; Galula & Nagl, 1964). This means that, depending on the dynamics of the conflict, insurgent groups can go back and forth between the three different phases of Mao's strategy. As a result, the defeat of the insurgency in a particular phase is often non-enduring. The active phase of the insurgency might even differ between different areas within a given nation or region if the organization seeks to expand geographically, as with JNIM in the Sahara-Sahel region. The developments in Mali before, during and after Operation Serval are a striking example of an insurgency switching between phases. Launched in 2013 to counter the advances of different Islamist militant groups, Operation Serval effectively pushed back the insurgents to the *Adrar des Ifoghas* and restored government control over the territories that the insurgents previously governed. However, despite the initial success of Operation Serval, the insurgency in Mali did not disappear entirely but rather regressed to a previous phase, most notably the strategic defense phase. In this phase, insurgent groups resorted to hit-and-run tactics, asymmetric warfare, and attacks on soft targets, thereby maintaining pressure on the government forces and prolonging the conflict. This regression highlights the insurgency's complex and dynamic nature, where gains made in one phase may not necessarily translate into enduring success, necessitating sustained and adaptive counterinsurgency efforts to address the evolving or regressing threats effectively.

Additionally, the progression through the phases of an insurgency is not strictly time-driven but rather conditions-based (Mackinlay, 2009). This reflects the complex interplay of political, societal, economic, and security dynamics within the insurgency. While Mao's three-phase model provides a theoretical framework for understanding the progress of an insurgency, the transition from one phase to another is largely dependent upon prevailing conditions and contextual factors rather than prescribed along a predefined timeline. Insurgencies evolve in response to changing circumstances, which can be shifts in government policies, fluctuations in public sentiment, external influences like the degree of outside support for belligerents, and the resilience of the insurgent's organizational structure. Thus, the duration and intensity of each phase can vary widely depending on the interplay of these factors. For instance, a resilient government response or effective counterinsurgency measures may prolong the initial phases of an insurgency, while systemic grievances or the termination of outside support to a government may accelerate the insurgency's progression to its final stage. Therefore, understanding the conditions that drive the transgression through insurgency phases is essential for designing tailored and effective counterinsurgency strategies that address root causes, exploit vulnerabilities, and disrupt insurgent activities.

Furthermore, it is critical to realize that insurgencies can deviate from the linear progression outlined in Mao's three-phase model (Kilcullen, 2010). It is essential to recognize that insurgent movements may skip phases or adapt their strategies based on changing circumstances and contextual factors. In particular, factors such as outside support, leadership dynamics, and the political and socio-economic environment can significantly influence the progress of an insurgency, potentially allowing it to bypass certain phases altogether (Jones, 2016). Both the phases of Mao and the ability of insurgencies to deviate from a linear progression make it increasingly harder to develop a codebook which is the best possible approximation of reality. Simultaneously, this also indicates that variables delineated according to Mao's phases might be less applicable when used in an analysis of a different NSAG (e.g. ISGS, another insurgency active in the Sahel region connected to the Islamic State). The usefulness of the model is therefore also reflected in the ability of the analyst to correctly define variables for the NSAG to be examined.

9.4 Conclusion

Using a machine learning-based model as decision support in counterinsurgency offers a promising approach to assess the progression of an insurgency, potentially aiding in the identification of transgressions through phases. By analyzing vast amounts of data, including historical records, machine learning algorithms can help identify patterns, trends, and correlations that may indicate shifts in the insurgents', intermediate objectives, capabilities and the resulting tactics. By defining the right variables and being able to generate predictive analysis, the model presented in this book can help anticipate the direction in which the insurgency is moving, whether it is intensifying, evolving, or regressing. Additionally, by integrating features that capture the key indicators of Mao's insurgency phases, such as levels of violence, territorial control, and popular support, the model can provide valuable insights into the progression of the insurgency and identify potential deviations from expected trajectories. Combining the predictive capabilities of machine learning with real-time intelligence and contextual information allows for a comprehensive qualitative analysis by SMEs. The result should enable security forces and policymakers to understand the dynamics of insurgencies better, anticipate emerging threats, and develop a proactive strategy to counter and defeat that of the insurgency. The TP-rules based model presented in this book forms a starting point from which to develop such decision support tools in counter insurgency.

As a precaution, it should be stated that insurgencies may adopt radical and expedited methods to achieve their objectives rather than Mao's protracted popular war. In the case of a coup, insurgents may seek to seize political power through forceful overthrow of the government. Consequently, counterinsurgents and especially analysts must remain open to recognizing the diversity of strategies insurgent movements may adopt and adjust the machine learning models accordingly. TP-rules, if based upon the right variables as defined by SME's that capture coup dynamics or guerilla warfare in an appropriate sense, can help in understanding and countering such activities. In general, by embracing flexibility and incorporating diverse perspectives into machine learning models, counterinsurgents can enhance their ability to anticipate, understand, and effectively counter insurgent movements by defeating their strategy.

References

Farrell, T. (2018). Unbeatable: Social resources, military adaptation, and the Afghan Taliban. *Texas National Security Review, 1*(3), 58–75.
Galula, D., & Nagl, J. A. (1964). *Counterinsurgency warfare: Theory and practice.* Praeger Security International. 2006.
Jones, S. G. (2016). *Waging insurgent warfare*. Oxford University Press.
Kaufmann, C. (1996). Possible and impossible solutions to ethnic civil wars. *International Security, 20*(4), 136–175.
Kilcullen, D. (2010). *Counterinsurgency*. Oxford University Press.
Mackinlay, J. (2009). *The insurgent archipelago: From Mao to Bin Laden*. Hurst.
Subrahmanian, V. S., Mannes, A., Sliva, A., Shakarian, J., & Dickerson, J. P. (2013). *Computational analysis of terrorist groups: Lashkar-e-Taiba*. Springer.
Tzu, S. (2000). *The art of war*. Translated by Lionel Giles. Allandale Online Publishing.
Waltz, E. (2014). *Quantitative intelligence analysis: Applied analytic models, simulations, and games*. Rowman & Littlefield.

Appendices

Appendix A: All TP-Rules

Appendix A contains all of the TP-Rules that have been written up in this book. Table A.1 contains the TP-Rules used to predict attacks.

Table A.1 TP-Rules used to predict attacks

Attack type	Condition A	Condition B	Condition C	Confidence	Offset
Abductions (release)	Com_Mge_ Campaign	Env_Gov_Intl_ Military Aid		0,54902	2
Abductions (release)	Env_GovSF_ Execution	not_Env_GovSF_ State of Emergency		0,71053	1
Abductions (release)	Rel_Intl_Travel Ban	not_Rel_Gov_ Raid		0,55319	6
Abductions	Rel_Intl_Travel Ban			0,91667	3
Abductions	Rel_Intl_Freeze Asset			0,89796	3
Abductions	Rel_Intl_Freeze Asset	not_Rel_Gov_ Raid		0,91489	1
Abductions	Rel_Intl_Travel Ban	not_Intra-Org Conflict w/o specifica		0,91489	1
Abductions	Rel_Intl_Freeze Asset	not_Env_GovSF_ Desertion		0,93333	3
Hit & Run attacks	Env_Gov_Intl_ Military Aid			0,64423	6
Hit & Run attacks	Env_Gov_Intl_ Military Aid	not_Env_GovSF_ Sexual Violence		0,66327	6
Hit & Run attacks	Env_Gov_Intl_ Military Aid	not_Lead_Att_ Fractious		0,65049	5
Attacks on Public Sites	Rel_Intl_Travel Ban			0,8125	4
Attacks on Public Sites	Rel_Intl_Freeze Asset			0,79592	4
Attacks on Public Sites	Rel_Intl_Freeze Asset	not_Rel_Neg_ Current Direct Negotiations		0,8125	5
Attacks on Public Sites	Rel_Intl_Travel Ban	not_Lead_Leg_ Election		0,80851	3
Attacks on Public Sites	Rel_Intl_Travel Ban	not_Com_Med_ News Media/ Periodicals		0,80851	3
Attacks on Public Sites	Rel_Intl_Freeze Asset	not_Lead_Att_ Arrest		0,8125	4
Attacks on Public Sites	Rel_Intl_Travel Ban	not_Lead_Att_ Deceased		0,85	3
Attacks on Public Sites	Rel_Intl_Freeze Asset	not_Rel_Gov_ Arrest Warrant		0,8125	4
Attacks on Public Sites	Rel_Intl_Travel Ban	not_Com_Mge_ Solidarity		0,79545	6
Attacks on Public Sites	Rel_Intl_Travel Ban	not_Rel_Gov_ Raid		0,82979	6

(continued)

Appendices

Table A.1 (continued)

Attack type	Condition A	Condition B	Condition C	Confidence	Offset
Targeting of civilians	Rel_Intl_Travel Ban			0,97917	5
Targeting of civilians	Rel_Intl_Freeze Asset			0,95918	4
Targeting of civilians	Rel_Intl_Freeze Asset	not_Lead_Att_ Arrest		0,97917	4
Targeting of civilians	Rel_Intl_Freeze Asset	not_Rel_Gov_ Arrest Warrant		0,97917	4
Targeting of civilians	Rel_Intl_Freeze Asset	not_Rel_Gov_ Raid		0,97872	4
Targeting of civilians	Rel_Intl_Freeze Asset	Env_Gov_Intl_ Military Aid		0,97778	4
Targeting of Public Sites	Com_Add_ Addressee w/o specifica	not_Env_Gov_ Intl_Border Closure		0,61538	4
Targeting of Public Sites	Com_Add_ Addressee w/o specifica	not_Group_Rel_ Unspecified Support given		0,64789	5
Absence of targeting of Security Forces	not_Rel_Intl_ Travel Ban	not_Com_Mge_ Strategy		0,73438	4
Absence of targeting of Security Forces	not_Com_Mge_ Campaign	not_Rel_Intl_ Freeze Asset		0,80702	1
Absence of targeting of Security Forces	not_Com_Mge_ Claim of Responsibility	not_Rel_Intl_ Freeze Asset		0,81481	1
Absence of targeting of Security Forces	not_Env_GovSF_ Execution	not_Env_GovSF_ State of Emergency		0,81481	2
Absence of targeting of Security Forces	not_Com_Mge_ Campaign	not_Env_GovSF_ State of Emergency		0,78723	2
Absence of targeting of Security Forces	not_Com_Mge_ Claim of Responsibility	not_Rel_Intl_ Freeze Asset	not_Env_ GovSF_ Sexual Violence	0,77551	2
Absence of targeting of Security Installations	not_Env_GovSF_ Execution	not_Env_GovSF_ State of Emergency		0,77778	2
Absence of targeting of Security Installations	not_Com_Mge_ Campaign	not_Env_GovSF_ State of Emergency		0,74468	5

Appendix B: Data Collection

The data we compiled spans from January 2011 to the end of 2022. In our dataset, we denote the occurrence of an attack or event in a given month with a 1, while the absence of such an event is indicated by a 0. Our data comes from various sources and databases. Specifically, for information on attacks and the targeting of individuals or sites, we relied on the Armed Conflict Location & Event Data Project (ACLED) database. Regarding communication specifics of JNIM, we predominantly utilized the services of BBC Monitoring, which tracks, translates, analyzes, and summarizes global media with expertise in JNIM and the Sahel. Lexisnexis, an academic news service, was employed for cross-referencing data when necessary or to fill gaps in situations where the aforementioned databases lacked information for certain variables. Table B.1 displays all the recorded variables, based on a codebook initially assembled by Jana Shakarian for the first author's previous study of Lashkar-e-Taiba, the terrorist group responsible for the notorious 2008 Mumbai attacks.

Table B.1 Definitions of variables

Variable	Explanation
Act_Abd_Abduction General	Did JNIM abduct people during the month in question?
Act_Abd_Release General	Did JNIM release abducted people during the month in question?
Act_Alt_Assassination	Did JNIM assassinate people during the month in question?
Act_Alt_Member Kill	Did members of JNIM reportedly kill other members of JNIM?
Act_Alt_Sexual Violence	Did JNIM commit acts of sexual violence during the month in question?
Act_Armed Clashes – Group's Casualties	During an armed clash, were casualties of JNIM reported?
Act_Armed Clashes – Security Forces Casualties	During an armed clash, were casualties of Security Forces reported?
Act_Armed Clashes – Unspecified Casualties	During an armed clash, were unspecified casualties reported?
Act_Ars_Arson General	Did JNIM reportedly use arson in an attack?
Act_Attack w/o specifica	Was there an attack without specific details?
Act_Attack_Attempted Attack	Did JNIM reportedly attempt an attack?
Act_Attack_Civilian Casualties	Were there civilian casualties?
Act_Attack_Government	Did JNIM attack the government?
Act_Attack_Hit & Run	Did JNIM commit a hit and run attack?
Act_Attack_Production Site	Did JNIM reportedly attack a production site?
Act_Attack_Public Site	Did JNIM attack a public site?
Act_Attack_School	Did JNIM attack a school?
Act_Bomb_Attempted Bombing	Did JNIM attempt a bombing?
Act_Bomb_Bombing General	Did JNIM use a non-suicide bombing?

(continued)

Table B.1 (continued)

Variable	Explanation
Act_Bomb_Suicide Bomb	Did JNIM reportedly commit a suicide bombing?
Act_Bus_Arms Trafficking	Did JNIM reportedly engage in arms trafficking?
Act_Bus_Extortion	Did JNIM reportedly engage in extortion?
Act_Bus_Human Trafficking	Did JNIM reportedly engage in human trafficking?
Act_Bus_Looting	Did JNIM reportedly engage in looting?
Act_Bus_Robbery w/o specifica	Did JNIM commit unspecified robberies?
Act_Bus_Sexual Exploitation	Did JNIM reportedly exploit anyone sexually?
Act_Mnp_Technical Manipulation w/o specifica	Did JNIM commit unspecified sabotages?
Act_S&C_Public Infrastructure	Did JNIM seize public infrastructure?
Act_S&C_Security Force's Structure	Did JNIM seize control of a structure belonging to security forces?
Act_S&C_Symbolic Site	Did JNIM seize control of a symbolic site?
Act_S&C_Territory	Did JNIM seize territory?
Act_Z_Civ_Belief	Did JNIM target civilians for their beliefs?
Act_Z_Civ_Civilian w/o specifica	Did JNIM target civilians without a specific reason?
Act_Z_Civ_Civilians Indiscriminate	Did JNIM target civilians indiscriminately?
Act_Z_Civ_Political Orientation	Did JNIM target civilians for their political orientations?
Act_Z_Prof_Government Official	Did JNIM target government officials?
Act_Z_Prof_School Teacher	Did JNIM target teachers?
Act_Z_Prof_Security Force	Did JNIM target security forces?
Act_Z_Prof_Production Site Personnel	Did JNIM target production site personnel?
Act_Z_Struc_Gov Building	Did JNIM target government buildings?
Act_Z_Struc_Security Installation	Did JNIM target security installations?
Act_Z_Struc_Production Site	Did JNIM target production sites?
Act_Z_Struc_Public Site	Did JNIM target public sites?
Act_Z_Struc_Public Transportation	Did JNIM target public transportations?
Act_Z_Struc_Structure w/o specifica	Did JNIM target an unspecified structure?
Act_Z_Struc_Symbolic Site	Did JNIM target a symbolic site?
Act_t_Election Day	Did JNIM attack on election day?
Com_Add_Addressee w/o specifica	Did JNIM communicate without a specific addressee?
Com_Add_Media	Did JNIM's communications address the media?
Com_Add_Government	Did JNIM's communications address the government?
Com_Add_Public	Did JNIM's communications address the public?
Com_Add_Security Forces	Did JNIM's communications address security forces?
Com_Add_NSAG	Did JNIM's communications address non-state actor groups?
Com_Med_E-mail	Was JNIM's communications medium email?

(continued)

Table B.1 (continued)

Variable	Explanation
Com_Med_Medium w/o specifica	Was JNIM's communications medium not specified?
Com_Med_News Media/Periodicals	Was JNIM's communications medium news media?
Com_Med_YouTube	Was JNIM's communications medium YouTube?
Com_Mge_Abusive Language	Did JNIM's communications contain abusive language?
Com_Mge_Aspirations & Objectives	Did JNIM's communications contain aspirations and objectives?
Com_Mge_Call for Recruits	Did JNIM's communications contain a call for recruits?
Com_Mge_Call for Violence	Did JNIM's communications contain a call for violence?
Com_Mge_Campaign	Did JNIM's communications contain a campaign?
Com_Mge_Claim of Responsibility	Did JNIM's communications contain a claim of responsibility?
Com_Mge_Declared Enemy	Did JNIM's communications establish their enemy?
Com_Mge_Justification of Violence	Did JNIM's communications contain a justification of violence?
Com_Mge_Solidarity	Did JNIM's communications contain a message of solidarity?
Com_Mge_Strategy	Did JNIM's communications contain a strategy?
Com_Mge_Threat	Did JNIM's communications contain a threat?
Com_Mge_Hostage	Did JNIM's communications mention a hostage?
Env_GovSF_Curfew	Did National Security Forces enforce a curfew?
Env_GovSF_Desertion	Did the National Security Forces desert?
Env_GovSF_Execution	Did National Security Forces execute civilians?
Env_GovSF_Forceful Resettlement	Did National Security Forces forcibly resettle people?
Env_GovSF_Remuneration of Security Forces	Did National Security Forces get paid?
Env_GovSF_Sexual Violence	Did National Security Forces commit acts of sexual violence?
Env_GovSF_State of Emergency	Did National Security Forces enforce a state of emergency?
Env_Gov_Intl_Allegation of Human Rights Abuse	Was the National government accused of human rights abuses?
Env_Gov_Intl_Allegation of War Crime	Was the National government accused of war crimes?
Env_Gov_Intl_Border Closure	Did the National government close the border?
Env_Gov_Intl_Foreign Aid	Did the National government receive foreign aid?
Env_Gov_Intl_Sanction, Resolution	Was the National government sanctioned?
Env_Gov_Intl_Travel Ban	Did the National government institute a travel ban?

(continued)

Table B.1 (continued)

Variable	Explanation
Env_Gov_Intl_Military Aid	Did the National government receive foreign military aid?
Env_Gov_Leg_Election	Was the National government elected legitimately?
Group_Basics_A.K.A	Did JNIM go by another name?
Group_Basics_Organization Strength	Were there reports on the group's strength?
Intra-Org Conflict General	Was there conflict within the JNIM organization?
Lead_Att_Academic	Was the leadership of JNIM trained or educated?
Lead_Att_Arrest	Was the leadership of JNIM arrested?
Lead_Att_Deceased	Was the leadership of JNIM killed?
Lead_Att_Fractious	Was the leadership of JNIM split/fractured?
Lead_Leg_Election	Was the leadership of JNIM elected?
Lead_Leg_External Installation	Was the leadership of JNIM installed by an external group?
Memb_Childsoldier	Were members of JNIM children?
Memb_Forceful Recruitment	Were members of JNIM recruited using force?
Memb_Foreigner	Were members of JNIM foreigners?
Memb_Socio-Economical Status	Were members of JNIM of a lower socio-economic status?
Group_Rel_Financial Support given	Did JNIM give financial support to another NSAG?
Group_Rel_Material Support given	Did JNIM give material support to another NSAG?
Group_Rel_Military Support given	Did JNIM give military support to another NSAG?
Group_Rel_Unspecified Support given	Did JNIM give unspecified support to another NSAG?
A&O_Ego_Profit	Did JNIM aim to make money?
A&O_Gov_ID_Non-Religious Rule	Did JNIM aim to install a non-religious government?
A&O_Gov_ID_Religious Rule	Did JNIM aim to install a religious government?
A&O_Gov_Political Aspirations & Objectives w/o specifica	Did JNIM have unspecified political objectives?
A&O_Gov_Political Participation	Did JNIM aim to participate politically?
Rel_Alliance General affirmed/formed	Did JNIM form an alliance with another NSAG?
Rel_Conflict Non-Violent w/NSAG	Did JNIM engage in a non-violent conflict with another NSAG?
Rel_Surrender	Did members of JNIM surrender?
Rel_Gov_Amnesty	Did the National government give amnesty to members of JNIM?
Rel_Gov_Arrest	Did the National government arrest members of JNIM?
Rel_Gov_Arrest Release	Did the National government release arrested members of JNIM?
Rel_Gov_Arrest Warrant	Did the National government issue arrest warrants?
Rel_Gov_Expulsion	Did the National government expel members of JNIM?

(continued)

Table B.1 (continued)

Variable	Explanation
Rel_Gov_Presence Close-Down	Did the National government close down JNIM locations?
Rel_Gov_Raid	Did the National government raid JNIM locations?
Rel_Gov_Trial	Did the National government put members of JNIM on trial?
Rel_Intl_Allegation of Human Rights Abuse	Did international actors accuse JNIM of human rights abuses?
Rel_Intl_Allegation of War Crime	Did international actors accuse JNIM of war crimes?
Rel_Intl_Arrest Extradition	Did international actors arrest and extradite a member of JNIM?
Rel_Intl_Designated Terror Organization	Did international actors designate JNIM a terror organization?
Rel_Intl_Embargo	Did another country put an embargo on Mali?
Rel_Intl_Freeze Asset	Did another country freeze JNIM's assets?
Rel_Intl_Travel Ban	Did another country ban travel to or from Mali?
Rel_Neg_Current Direct Negotiations	Are negotiations between the National government and JNIM occurring?
Rel_Neg_Current Indirect/3rd Party Talks	Are negotiations between a third party and JNIM occurring?
Rel_Neg_End_Ceasefire	Was a ceasefire negotiated?
Rel_Neg_Group Refuse	Did JNIM refuse to negotiate?
Rel_Neg_Negotiations Planned	Are negotiations planned between JNIM and the National government?
Rel_Neg_State Refuse	Did the National government refuse to negotiate with JNIM?

Appendix C: Most Frequently Occurring Independent Variables

Appendix C contains a table of the variables most frequently used by the TP-Rules that we derived from our JNIM data.

Most used independent variables in TP-rules	
Variable	No. of times used
A travel ban was placed on the country where JNIM operates	13
Foreign states or international institutions froze asset(s) of individual member(s) of JNIM	13
The national government did not raid JNIM facilities and locations	3
JNIM discusses whether their campaign has been achieving the desired objectives	2
The government did not warrant arrests of members of JNIM	2
The Malinese, Nigerian or Burkinese government reportedly ordered execution(s) of de(	1
JNIM was not in direct negotiations with the nation with which it is in conflict	1
The leadership of JNIM was not elected	1
JNIM did not communicate through print media (magazines, journals for publishing articl	1
The top leaders of JNIM were not under arrest or imprisoned in the coded period	1
The top leaders of JNIM did not die in the period being coded	1
JNIM did not communicate a message of solidarity	1
The national government's security forces did not execute civilians	1

Appendix D: Network Visualizations

The evolving events corresponding to the various JNIM activities are presented as a time-evolving network parametrized by Δ, ε and β, as loosely defined below. Given time periods $t \in \{1,\ldots,T\}$, we define the time-evolving network as $G^T = (V_\Delta^T, \mathcal{E}^T_{\varepsilon,\beta})$, with $V_\Delta^T = \bigcup_{t=n-\Delta}^{n} V^t$ and $V^t \subset \{v_1, v_2, \ldots, v_n\}$ is the set of nodes present at time t, and $\mathcal{E}^T_{\varepsilon,\beta} = \bigcup_{t=n-\Delta}^{n} E^t$ is the set of undirected edges in $V_\Delta^T \times V_\Delta^T$ such that the geodesic distance between nodes is less than or equal to ε kilometer and nodes 'occur' within β days of each other. In other words, for every $(v_i, v_j) \in \mathcal{E}^T_{\varepsilon,\beta}$ it holds that $d(v_i, v_j) \leq \varepsilon$ and $|t_i - t_j| \leq \beta$, where $d(.,.)$ corresponds to geodesic distance (in [km]) between nodes and t_i equals the day the corresponding event v_i occurred. For every visualization we compute three standard network metrics: average betweenness, average clustering and density.[1] Averages are taken first over individual connected components values and second over connected components as a whole. All figures in this appendix were generated using Python's Matplotlib library, incorporating publicly available data sources.

D.1 No-Decay Visualizations (Figs. D.1, D.2, D.3, D.4, D.5, D.6, D.7, D.8, D.9, D.10, D.11, D.12, D.13, D.14 and D.15)

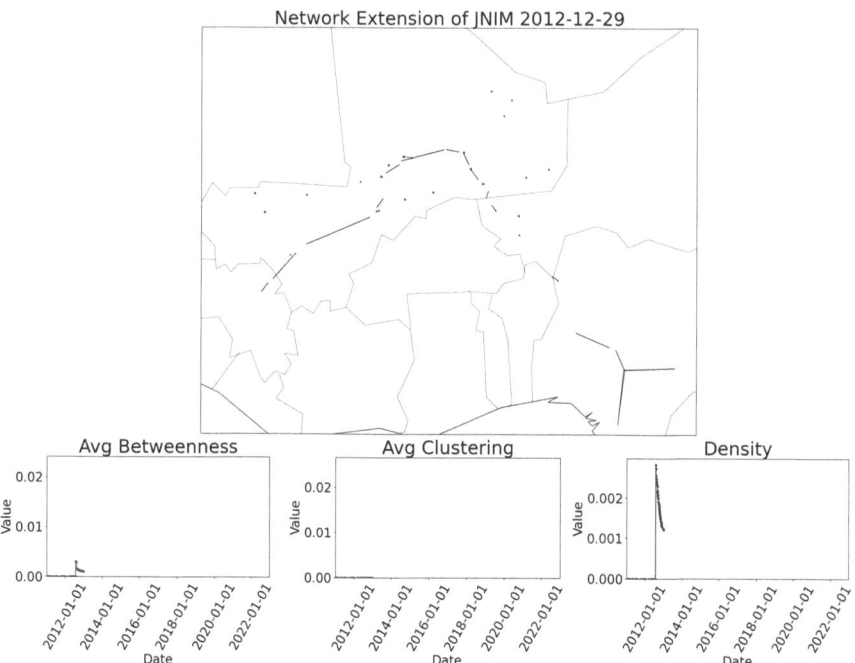

Fig. D.1 JNIM's time evolving event network of December 29, 2012. With $\Delta = \infty$, $\varepsilon = 40$ [km], $\beta = 30$ [days]

---

[1] See for instance Bloch, Francis, Matthew O. Jackson, and Pietro Tebaldi. "Centrality measures in networks." *Social Choice and Welfare* 61.2 (2023): 413-453.

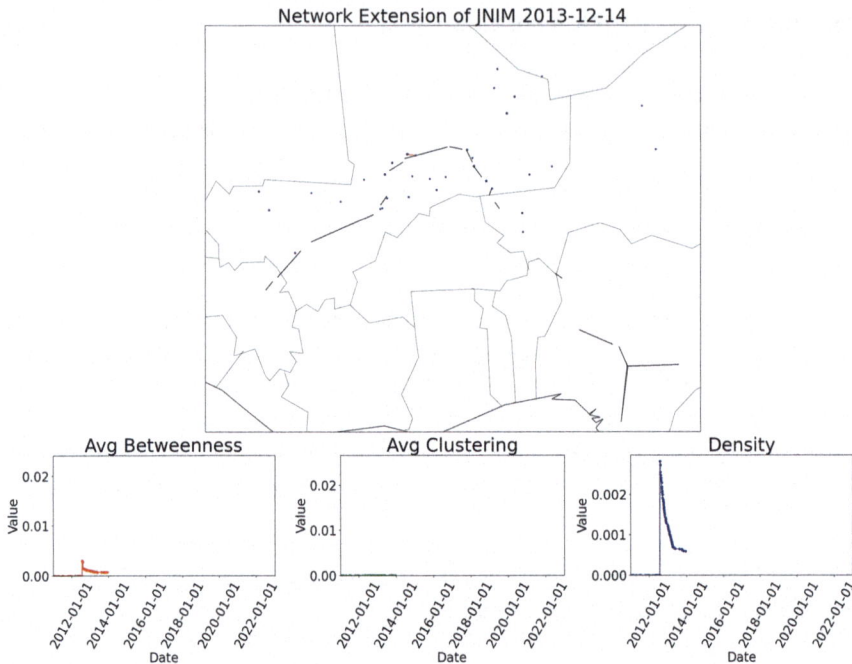

Fig. D.2 JNIM's time evolving event network of December 14, 2013. With $\Delta = \infty$, $\varepsilon = 40$ [km], $\beta = 30$ [days]

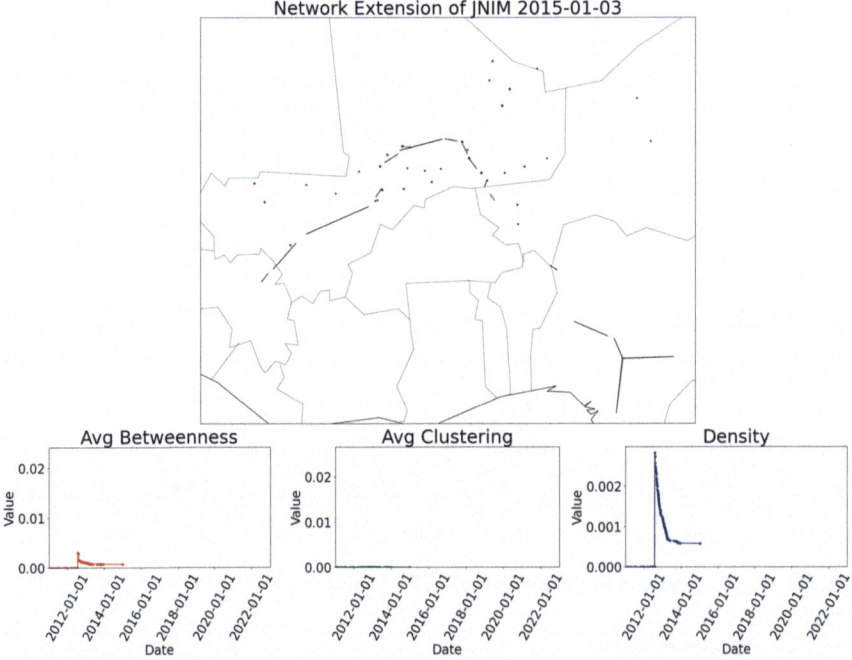

Fig. D.3 JNIM's time evolving event network of January 03, 2015. With $\Delta = \infty$, $\varepsilon = 40$ [km], $\beta = 30$ [days]

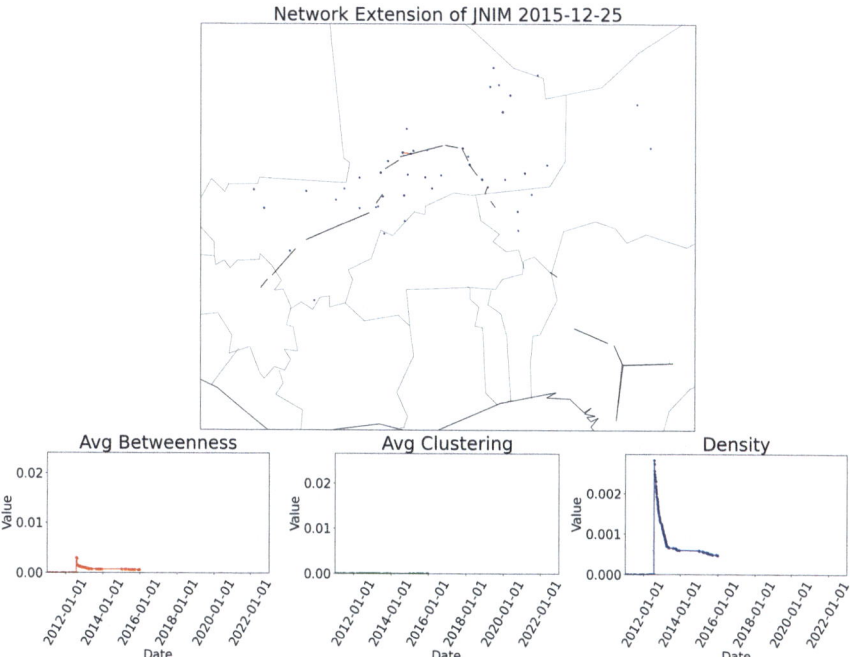

Fig. D.4 JNIM's time evolving event network of December 25, 2015. With $\Delta = \infty$, $\varepsilon = 40$ [km], $\beta = 30$ [days]

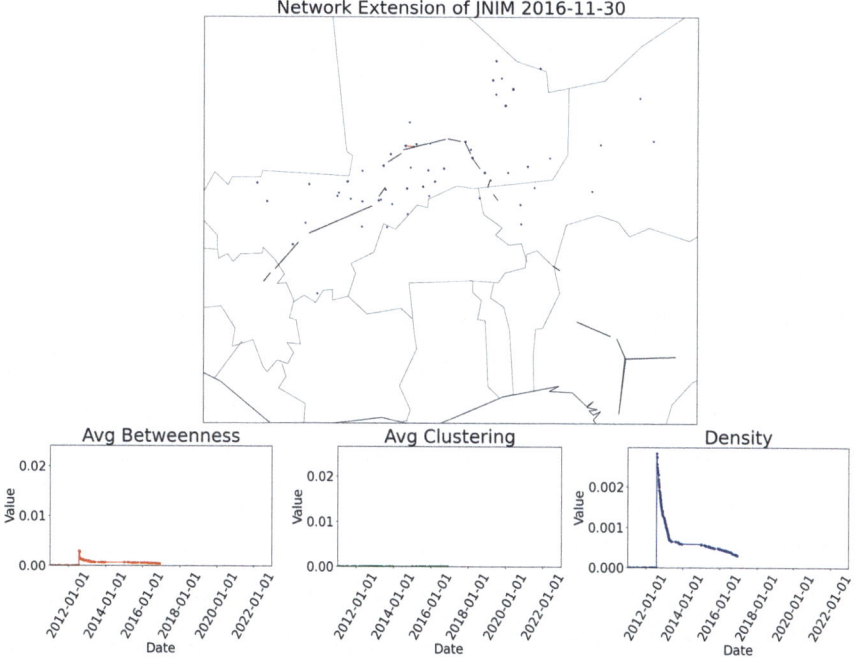

Fig. D.5 JNIM's time evolving event network of November 30, 2016. With $\Delta = \infty$, $\varepsilon = 40$ [km], $\beta = 30$ [days]

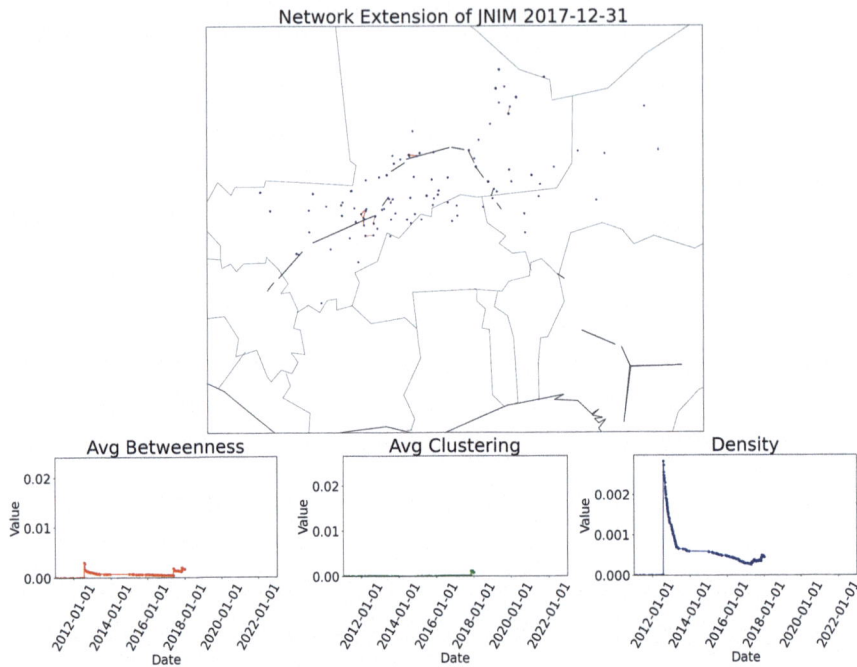

Fig. D.6 JNIM's time evolving event network of December 31, 2017. With $\Delta = \infty$, $\varepsilon = 40$ [km], $\beta = 30$ [days]

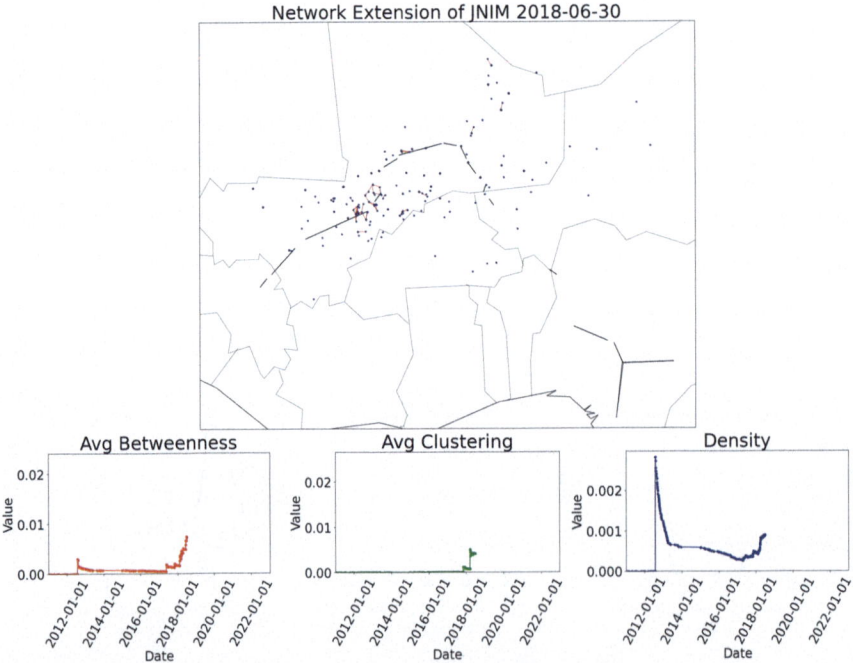

Fig. D.7 JNIM's time evolving event network of June 30, 2018. With $\Delta = \infty$, $\varepsilon = 40$ [km], $\beta = 30$ [days]

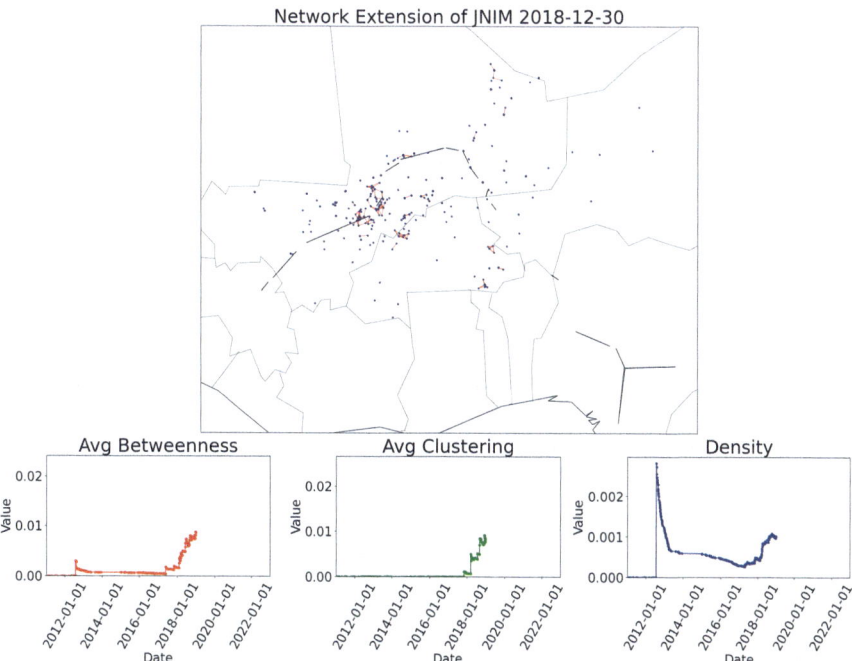

Fig. D.8 JNIM's time evolving event network of December 30, 2018. With $\Delta = \infty$, $\varepsilon = 40$ [km], $\beta = 30$ [days]

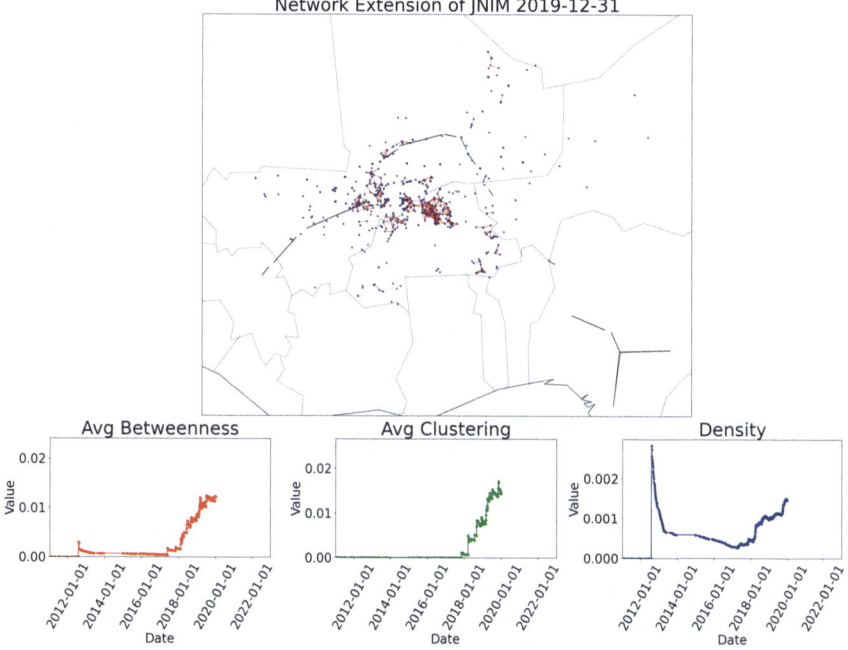

Fig. D.9 JNIM's time evolving event network of December 31, 2019. With $\Delta = \infty$, $\varepsilon = 40$ [km], $\beta = 30$ [days]

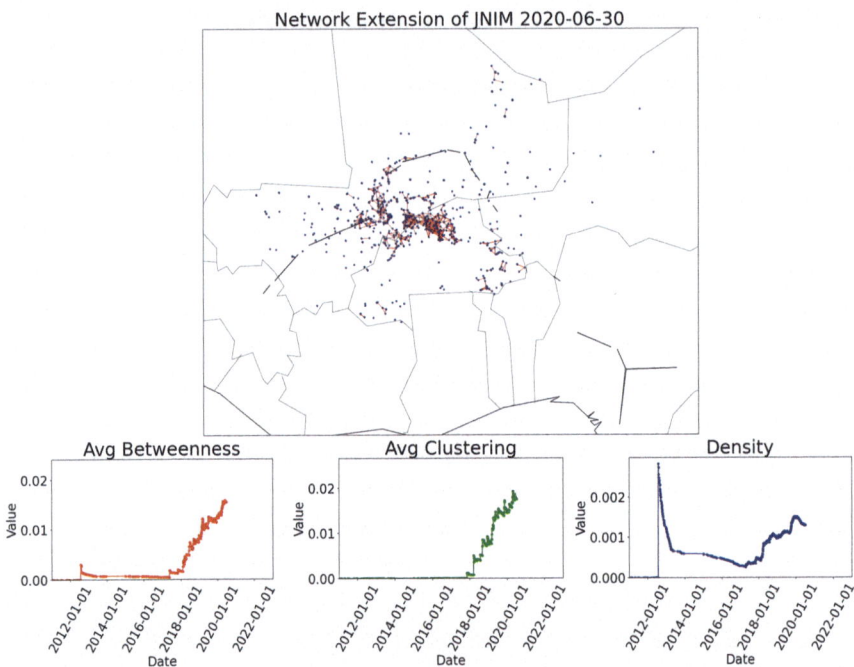

Fig. D.10 JNIM's time evolving event network of June 30, 2020. With $\Delta = \infty$, $\varepsilon = 40$ [km], $\beta = 30$ [days]

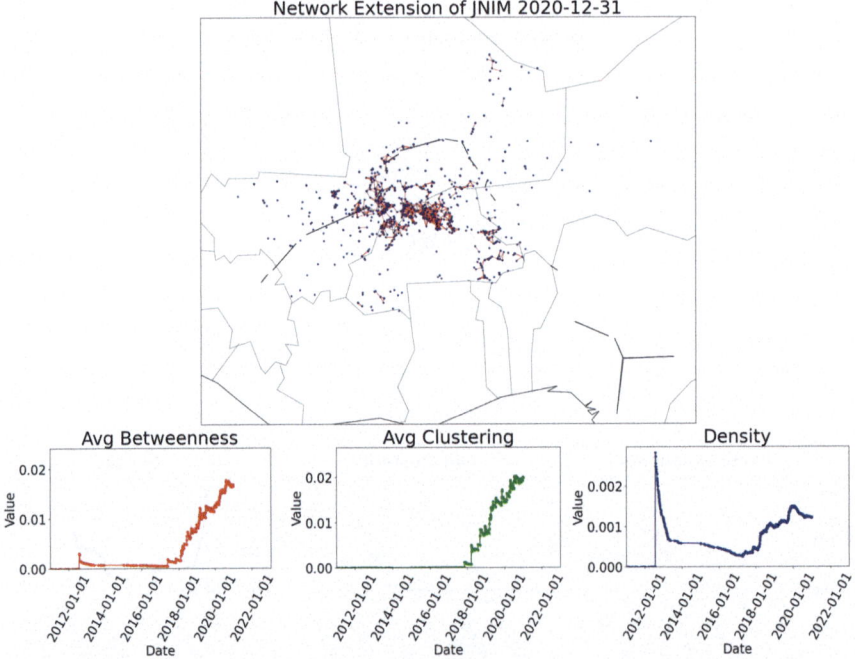

Fig. D.11 JNIM's time evolving event network of December 31, 2020. With $\Delta = \infty$, $\varepsilon = 40$ [km], $\beta = 30$ [days]

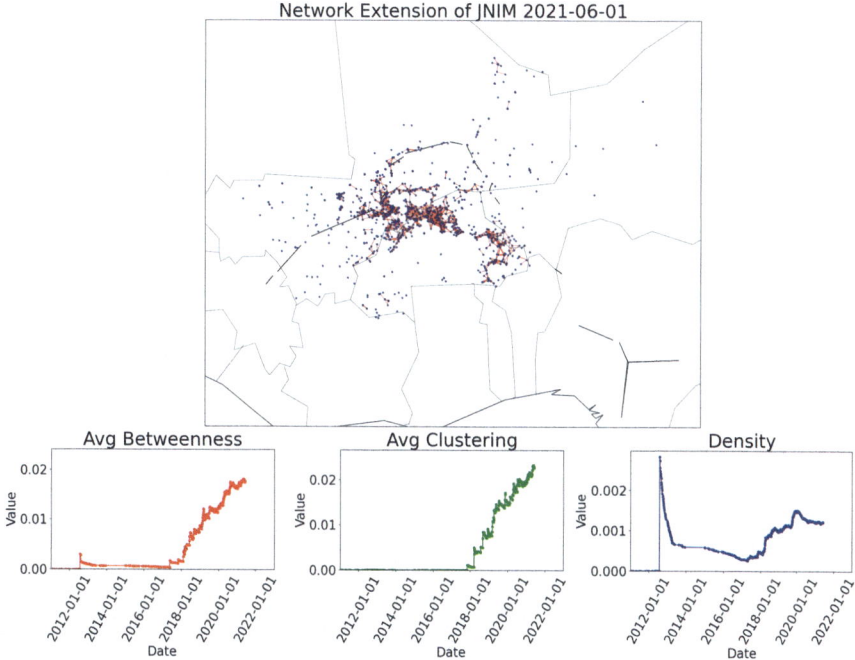

Fig. D.12 JNIM's time evolving event network of June 30, 2021. With $\Delta = \infty$, $\varepsilon = 40$ [km], $\beta = 30$ [days]

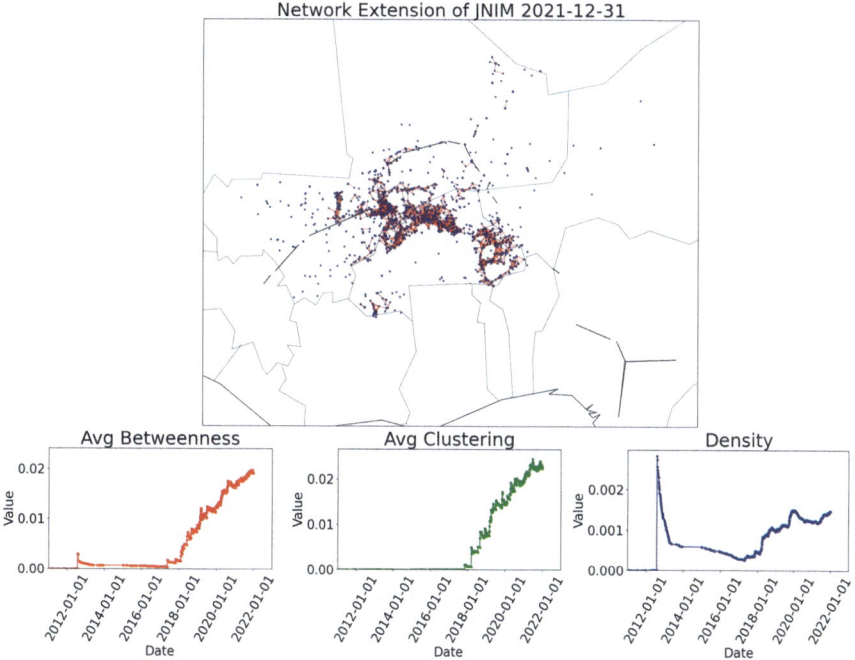

Fig. D.13 JNIM's time evolving event network of December 31, 2021. With $\Delta = \infty$, $\varepsilon = 40$ [km], $\beta = 30$ [days]

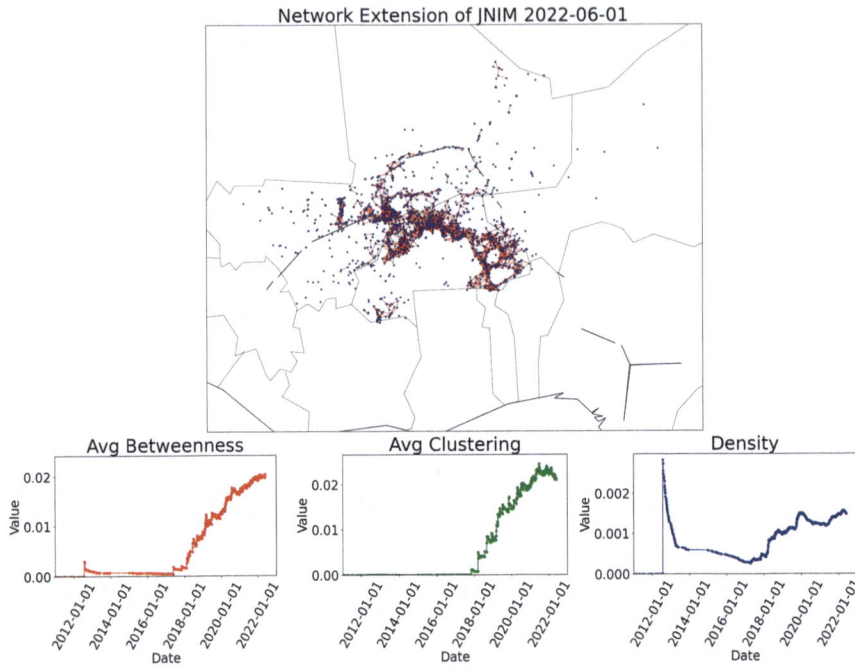

Fig. D.14 JNIM's time evolving event network of June 01, 2022. With $\Delta = \infty$, $\varepsilon = 40$ [km], $\beta = 30$ [days]

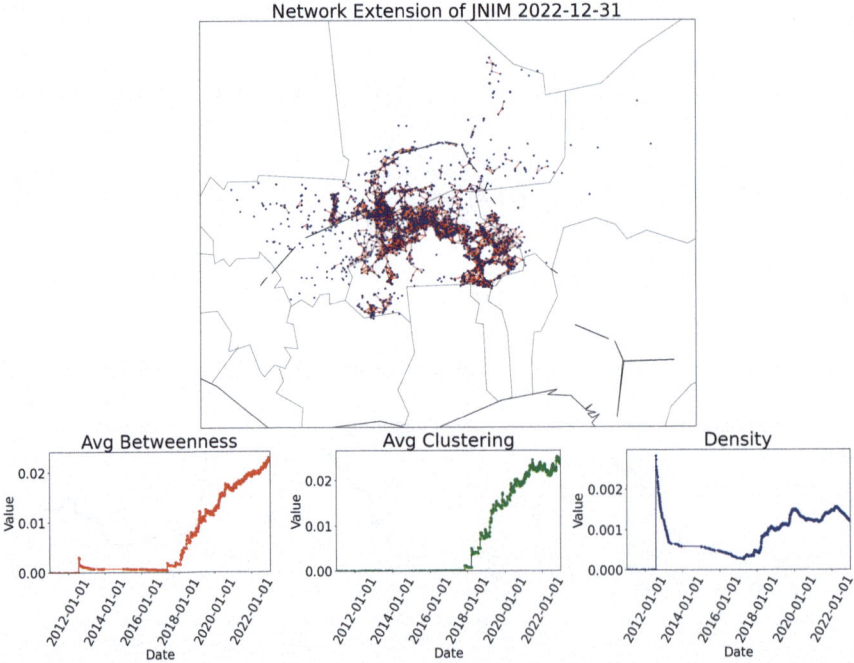

Fig. D.15 JNIM's time evolving event network of December 31, 2022. With $\Delta = \infty$, $\varepsilon = 40$ [km], $\beta = 30$ [days]

The manufacturer's authorised representative in the EU is Springer Nature Customer Service Centre GmbH, Europaplatz 3, 69115 Heidelberg, Germany. If you have any concerns regarding our products, please contact ProductSafety@springernature.com

Printed and bound by CPI Group (UK) Ltd, Croydon, CR0 4YY

26/03/2026

02078979-0005